# SIMPLY

## ARTIFICIAL
## INTELLIGENCE

**DK LONDON**

**Senior Editor** Chauney Dunford
**Senior Designer** Mark Cavanagh
**Editors** Daniel Bryne, Elizabeth Dowsett, Steve
Setford, Alison Sturgeon, Andrew Szudek
**Managing Editor** Gareth Jones
**Senior Managing Art Editor** Lee Griffiths
**Production Editor** Robert Dunn
**Senior Production Controller** Rachel Ng
**Jacket Design Development Manager**
Sophia M.T.T.
**Jacket Designer** Akiko Kato
**Associate Publishing Director** Liz Wheeler
**Art Director** Karen Self
**Publishing Director** Jonathan Metcalf

First published in Great Britain in 2023 by
Dorling Kindersley Limited
DK, One Embassy Gardens, 8 Viaduct Gardens,
London, SW11 7BW

A CIP catalogue record for this book
is available from the British Library.
ISBN: 978-0-2416-0728-2

Printed and bound in China

## For the curious
**www.dk.com**

## CONSULTANT
**Hilary Lamb** is an award-winning science and technology journalist, editor, and author. She studied physics at the University of Bristol and science communication at Imperial College London before spending five years as a staff magazine reporter. She has worked on previous DK titles, including *How Technology Works*, *The Physics Book*, and *Simply Quantum Physics*.

## EDITORIAL CONSULTANT
**Joel Levy** is a writer who specializes in science and the history of science. His writing explores both mainstream science and weird technology, and his books include *The Infinite Tortoise: The Curious Thought Experiments of History's Great Thinkers*, *Gothic Science: The Era of Ingenuity and the Making of Frankenstein*, and *Reality Ahead of Schedule: How Science Fiction Inspires Science Fact*.

## CONTRIBUTOR
**Dr Claire Quigley** is a computing scientist who has previously worked for the Universities of Cambridge and Glasgow. She has contributed to the creation of coding activities for the BBC and Virgin Media, and written for several previous DK titles, including *Help Your Kids with Computer Science*, *Computer Coding Python Games for Kids*, and *Computer Coding Python Projects for Kids*.

# CONTENTS

# STATISTICAL ARTIFICIAL INTELLIGENCE

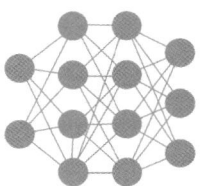

# USING ARTIFICIAL INTELLIGENCE

# PHILOSOPHY
OF **ARTIFICIAL INTELLIGENCE**

# LIVING WITH ARTIFICIAL INTELLIGENCE

# WHAT IS ARTIFICIAL INTELLIGENCE?

Artificial intelligence (AI) is intelligence demonstrated by machines – which in turn are known as "AIs". The history of AI dates back to the 1950s, when the first modern computers were built. The decades since then have seen waves of excitement and disillusionment, and a shift of focus from AIs based on formal logic (known as "classical" or "symbolic" AIs) to AIs based on data and statistics. Today, machine learning – the use of large data sets to train AI models, such as artificial neural networks, to perform tasks without being explicitly programmed to do so – dominates AI research. Using this approach, models can be taught to perform tasks quickly and expertly.

In popular culture, AIs are often depicted as being rivals of human intelligence – even as an existential threat. In reality, AI technologies tend to be limited in their applications – a long way from reaching the intelligence of a cat, let alone a human being. However, AI is a powerful tool when applied to specific problems, such as reading handwriting, recommending TV shows, or diagnosing medical conditions.

We use AIs every day without noticing it. However, as they take over ever more human tasks their prevalence raises urgent and complex questions about how we can ensure that AIs continue to serve the whole of humanity, and not just themselves or a powerful elite. Seeing machines perform tasks that were previously considered uniquely human, even creating art and music, challenges our most fundamental assumptions about what it means to be human. Our future with AI is uncertain, but it is one that scientists, engineers, mathematicians, philosophers, policymakers, and anyone else with an interest in humanity's future can help to shape.

# HISTORY
# ARTIFIC
# INTELLI

# O F

# I A L

# G E N C E

**Long before** AI became a practical possibility, the notion of a "living machine" existed in mythology, particularly in the tales of Ancient Greece and China. However, the idea was first taken seriously in the 18th century, when engineers created complex, self-propelled devices – or "automata". Meanwhile, philosophers pondered whether human thought could be simulated by manipulating symbols – an idea that led to the invention of the first programmable digital computers in the 1940s. By the end of the 1950s AI was a recognised field of study, and computers have grown more powerful ever since. This in turn has led to the creation of increasingly versatile AIs – although none that can be said to be "alive".

# AN IMITATION OF LIFE

An automaton is a machine that is able to operate on its own, following a sequence of programmed instructions. Historically, most automata were animated toys – often clockwork figures or animals, some of which were surprisingly lifelike. Animatronics, which are typically used to portray film or theme-park characters, are modern electronic automata.

In AI, the word "automaton" refers to a computer that can be programmed to perform a specific task, such as forecast the stock market or analyse customer behaviour. The latest AIs are highly sophisticated, and appear to have minds of their own. However, one has yet to be built that can control its own actions.

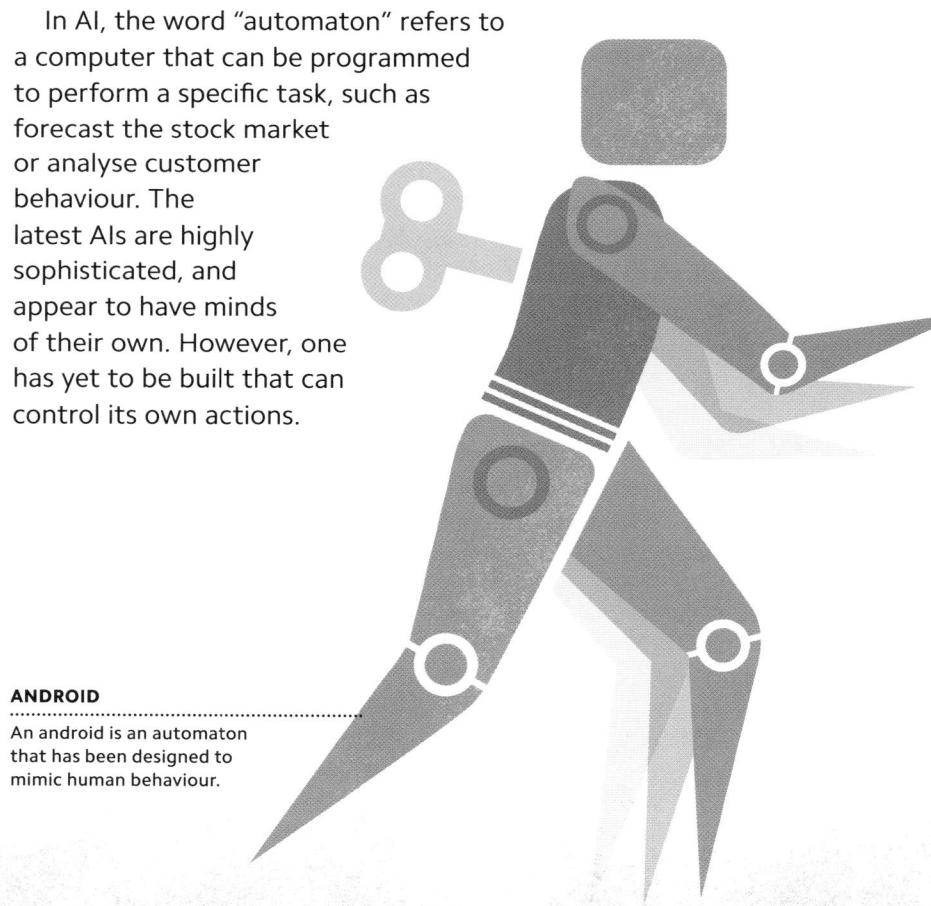

**ANDROID**

An android is an automaton that has been designed to mimic human behaviour.

LINGUISTIC INTELLIGENCE

SPATIAL INTELLIGENCE

# DEFINING INTELLIGENCE

ARTISTIC INTELLIGENCE

EMOTIONAL INTELLIGENCE

English mathematician Alan Turing (1912–54) devised a test that can be used to establish whether or not a machine has human-like intelligence (see pp.130–31). Originally, the Turing test focused on numerical intelligence (the ability to perform mathematical calculations). However, scientists now argue that since there are different kinds of intelligence (such as artistic and emotional intelligence), an AI must demonstrate each kind of intelligence for it to be considered the equivalent of a human being. Broadly speaking, there are eight kinds of intelligence, including sensory intelligence (the ability to interact with one's environment) and reflective intelligence (the ability to reflect upon and modify one's behaviour).

NUMERICAL INTELLIGENCE

PHYSICAL INTELLIGENCE

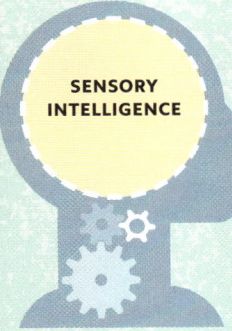

SENSORY INTELLIGENCE

REFLECTIVE INTELLIGENCE

"I know that I am intelligent, because I know that I know nothing."
Socrates

# THINKING = COMPUTING

The idea that all thinking, whether human or artificial, is a form of computing (see p.15) – specifically, a process of using algorithms to convert symbolic inputs into symbolic outputs (see p.36) – is known as "computationalism". Computationalists argue that the human brain is a computer, and that, one day, an AI should therefore be able to do anything that a brain can do. In other words, they claim, such an AI would not merely simulate thinking, it would have genuine, human-like consciousness.

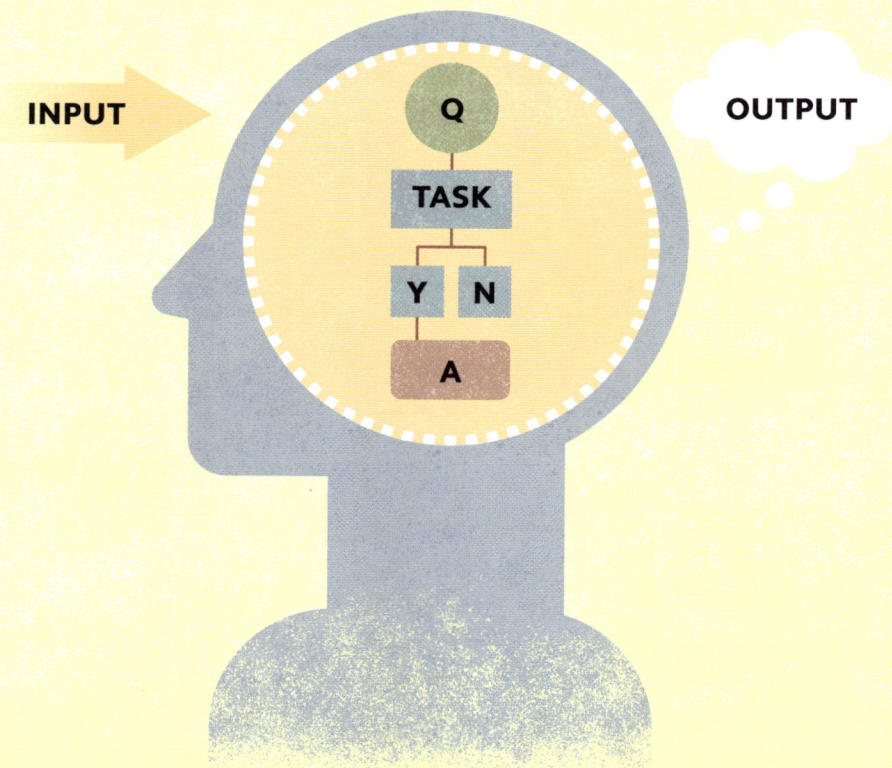

INPUT

Q

TASK

Y   N

A

OUTPUT

# ZEROS AND ONES

A binary code is a code that represents information, such as instructions, using only two numbers, or digits. The binary code most commonly used in computing features the numbers 0 and 1, each of which represents a "bit" of information. Any number can be converted into zeros and ones (for example, the decimal number 12 is 1100 in binary), as can any letter of any known alphabet. The two digits can also represent the two states of an electrical current – "on" or "off" – meaning that software translated into binary code can be read by a computer.

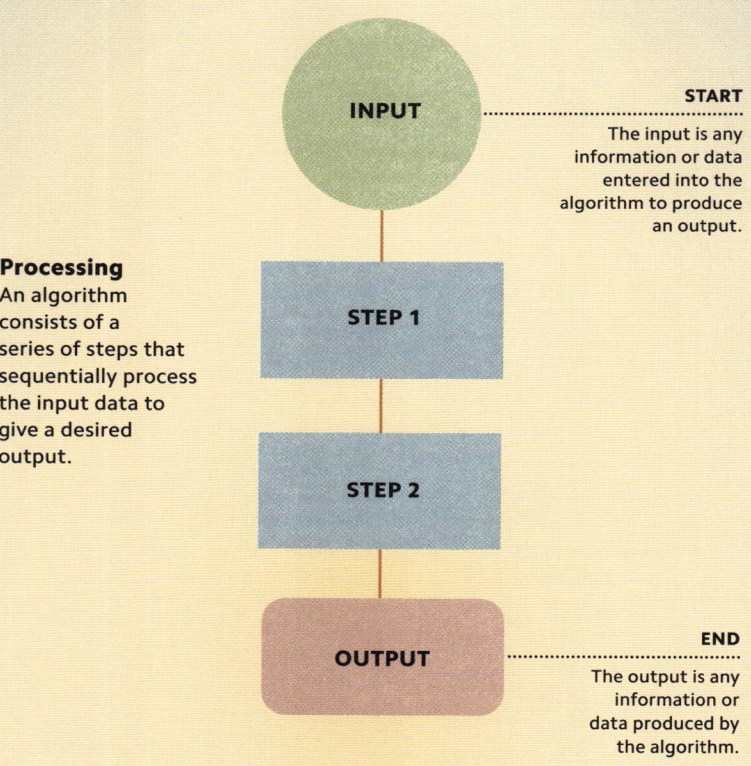

**START**

The input is any information or data entered into the algorithm to produce an output.

**Processing**
An algorithm consists of a series of steps that sequentially process the input data to give a desired output.

**END**

The output is any information or data produced by the algorithm.

# STEP BY STEP

An algorithm is a sequence of instructions for accomplishing a task. It takes an input, such as information or data, and processes it in a series of steps to produce a desired result, or output. The task or process can range from a simple calculation, or following a recipe to make a meal, to solving complex mathematical equations. An algorithm is an example of what mathematicians call an "effective method", which means it has a finite number of steps and produces a definite answer, or output.

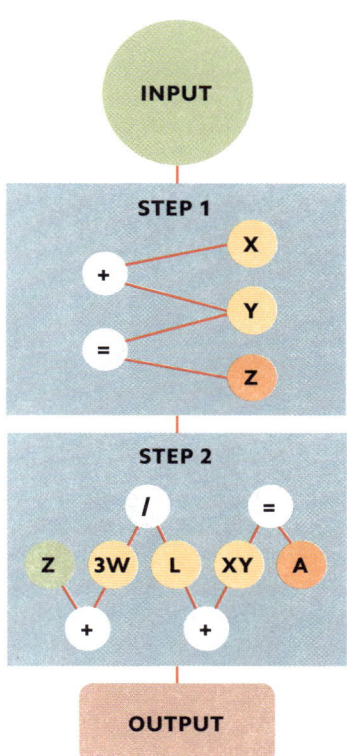

**Components of calculation**
Computations have an input and an output, multiple steps, and can vary from simple sums to complex equations.

# ALGORITHMS IN ACTION

A computation is a calculation that follows the steps of an algorithm (see opposite). The most straightforward example of computation is arithmetic calculation. For example, if you add a pair of three-digit numbers together in your head, you follow a series of steps, or an algorithm, to achieve this calculation. Computations use symbols to represent numbers, but symbols can represent almost anything else (see p.36). With the right symbols and the right algorithms, immensely complex computation becomes possible.

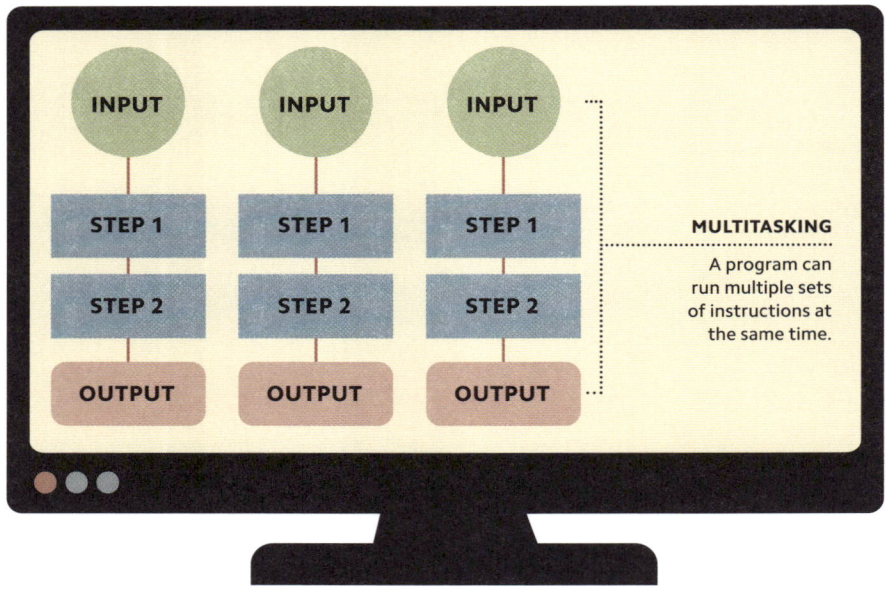

# INSTRUCTING COMPUTERS

A program is a sequence of instructions written in code that enables a computer to perform one or more tasks. Charles Babbage (see opposite) imagined the first program. He was inspired by the design of a certain silk loom, which had parts that moved up or down in response to a pattern of holes punched into a card. Babbage recognized that these holes could store instructions to operate the cogs and levers of a machine that he was designing: the "Analytical Engine". Modern computers work on the same principle, following sequences of instructions, which are usually written in binary code (see p.13).

# THE FIRST MECHANICAL COMPUTERS

In the 19th century, the complex work of producing numerical tables (used in navigation, warfare, and other fields) was performed by people known as "computers". To avoid mistakes caused through human error, English mathematician Charles Babbage (1791–1871) invented what he called the "Difference Engine" – a machine that could perform mathematical calculations mechanically. Babbage then designed the "Analytical Engine" – a general-purpose calculator that could be programmed using punched cards (see opposite), and had separate memory and processing units. Although it was never built, the Analytical Engine had many of the key features of modern computers (see p.22).

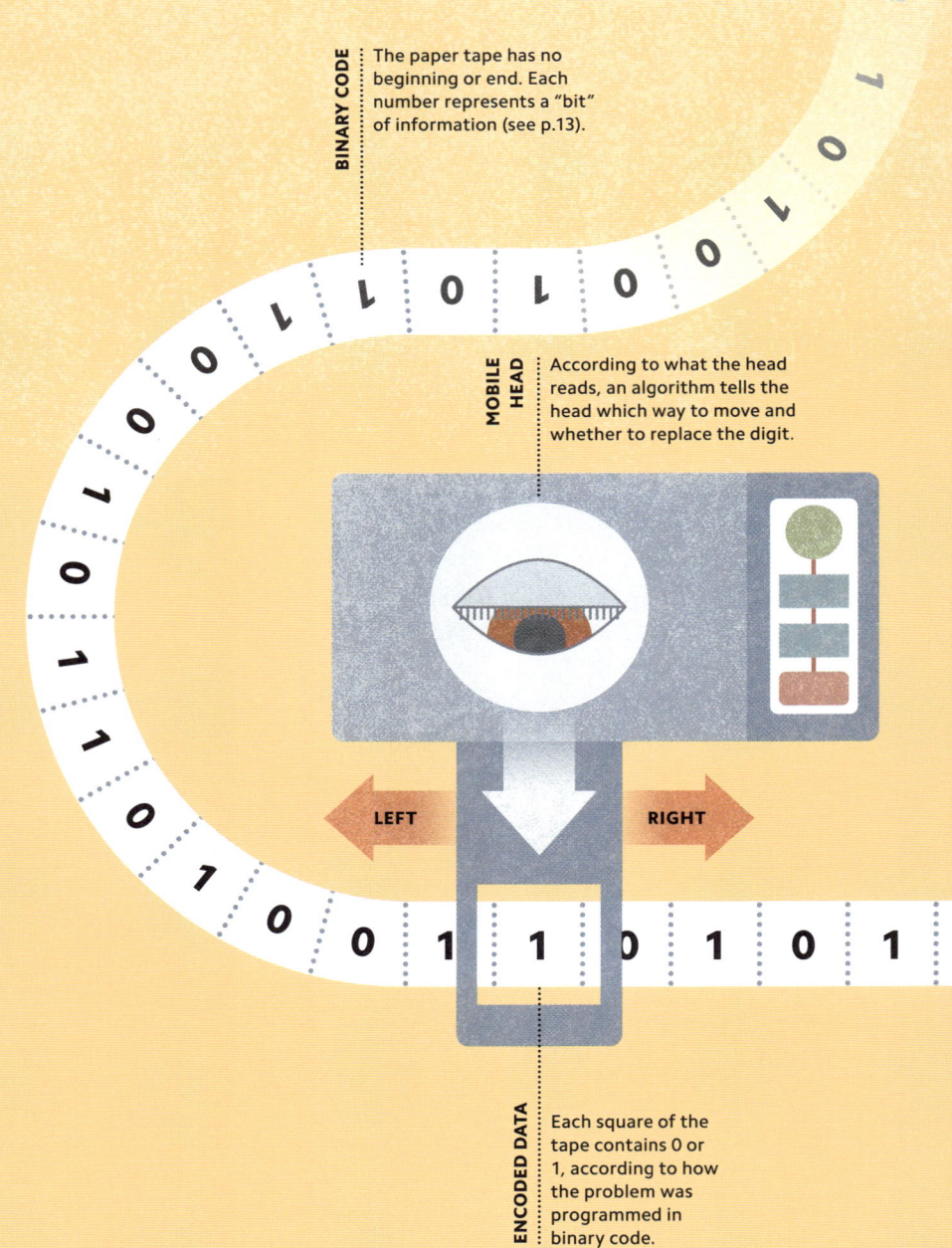

BINARY CODE

**BINARY CODE** The paper tape has no beginning or end. Each number represents a "bit" of information (see p.13).

**MOBILE HEAD** According to what the head reads, an algorithm tells the head which way to move and whether to replace the digit.

LEFT

RIGHT

**ENCODED DATA** Each square of the tape contains 0 or 1, according to how the problem was programmed in binary code.

# A THEORETICAL COMPUTER

In 1936, English mathematician Alan Turing (1912–54) proposed an imaginary machine that could solve any problem that could be made "computable" (see p.15). In other words, so long as the problem could be written using symbols and algorithms, and translated into binary code (see p.13), his machine could solve it. The device consisted of a head that moved over a tape marked with binary information. Although it was never built, Turing's Universal Machine sparked the computer revolution by proving that a machine could tackle any computable problem.

**1 0 1 1 1 0 0 0 1 0 1 1**

**Problem-solving machine**
A "read/write" head moves back and forth along a paper tape. Following instructions from an algorithm, it changes 1s to 0s, and vice versa, depending on what has come before.

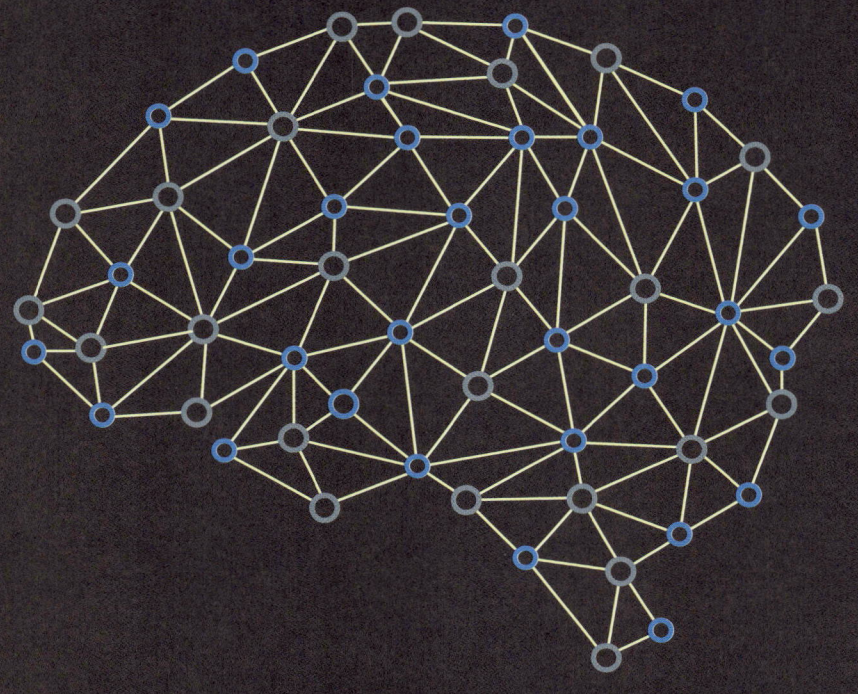

# AN ELECTRIC BRAIN

Alan Turing (see pp.18–19) had shown that a machine could carry out any computation (see p.15) with the right combination of symbols. In 1943, scientist Walter McCulloch (1898–1969) and mathematician Walter Pitts (1923–69) demonstrated that networks of units based on human nerve cells, or neurons, passing electric signals back and forth, could copy a Turing machine. They suggested that the brain might be a kind of living computer, meaning that a program that ran on the human brain might also run on an electric brain. This theory is known as the principle of "multiple realizability" (see p.136).

# ARTIFICIAL NEURONS

Each of the 86 billion neurons in the human brain is
effectively a tiny processor, receiving electrical signals (inputs)
from other neurons and sending out signals of its own
(outputs). McCulloch and Pitts (see opposite) realized that
neurons can act as logic gates – devices that can switch on and
off (see p.13), depending on the input. The scientists described
an imaginary neuron called a "threshold logic unit". This neuron
works by first adding the values of its inputs (signals from other
neurons) and then multiplying that value by a variable called
a "weight" (see p.78) – this is the strength of a connection
between neurons. If the input signals exceed a certain value
(see p.79), the neuron is triggered to send an output signal.
This triggering is called the "activation function".

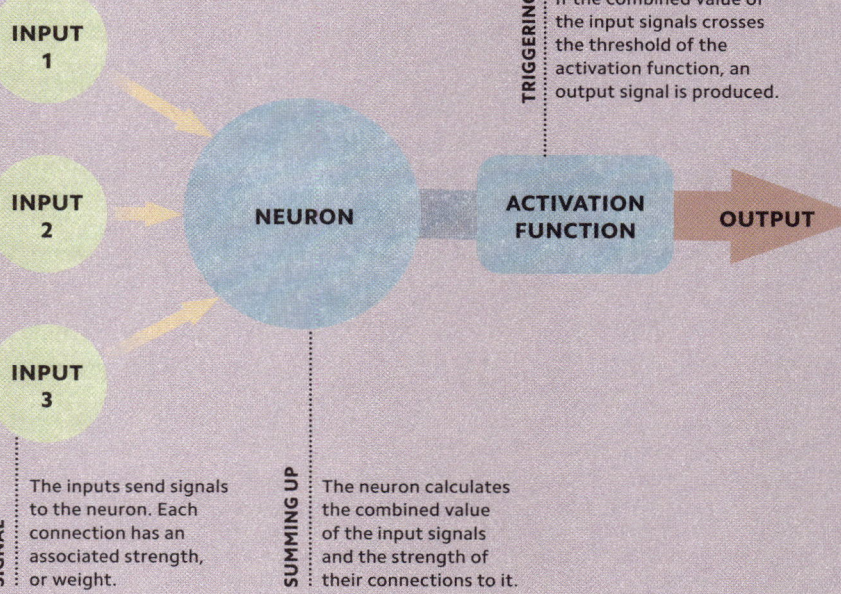

INPUT 1

INPUT 2

INPUT 3

NEURON

ACTIVATION FUNCTION

OUTPUT

**TRIGGERING**
If the combined value of
the input signals crosses
the threshold of the
activation function, an
output signal is produced.

**SIGNAL**
The inputs send signals
to the neuron. Each
connection has an
associated strength,
or weight.

**SUMMING UP**
The neuron calculates
the combined value
of the input signals
and the strength of
their connections to it.

# A PROGRAMMABLE COMPUTER

The Electronic Numerical Integrator and Computer (ENIAC) was an early electronic computing machine built in the US between 1943 and 1946. Made up of over 18,000 vacuum tubes (electronic components resembling light bulbs) and covering 167 sq m (1,800 sq ft), it calculated range tables (a list of the angles and elevation needed to hit a target) for artillery, giving its answers on paper punch cards. In just 20 seconds, it could complete a calculation that took people hours using electromechanical calculators. ENIAC was programmed by changing the arrangement of cables that plugged into it, which took days to complete. It was the first machine computer that could run different programs.

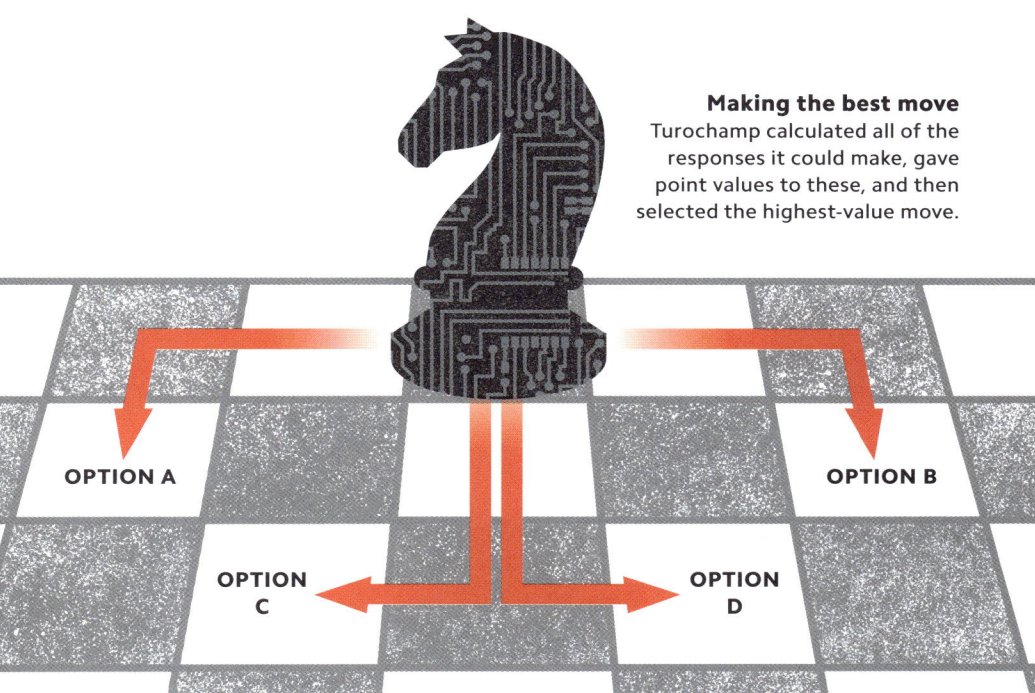

**Making the best move**
Turochamp calculated all of the responses it could make, gave point values to these, and then selected the highest-value move.

**OPTION A**

**OPTION B**

**OPTION C**

**OPTION D**

# A THEORETICAL PROGRAM

In 1948, Alan Turing (see pp.18–19) and mathematician David Champerowne (1912–2000) set out to prove that, with the right algorithm, a computer could play a game of chess. At the time, no electronic computer existed that could run such an algorithm, so Turing played the role of computer himself, performing each step of the algorithm on paper. "Turochamp", as they called it, was further proof that computers (whether human or artificial) could perform complex calculations without understanding what they were doing, but simply by following a set of instructions.

# A COMPUTING BLUEPRINT

John von Neumann (1903–57) was a Hungarian–American scientist involved in developing ENIAC (see p.22), the first programmable computer. He devised a model (see right) that established how the main components of modern-day computers are structured – known as von Neumann architecture. The major advancement was the use of a memory unit that contained both the programs (see p.16) and data (see p.32), making the machines quicker and easier to reprogram than existing ones. Information within the memory unit feeds into a central processing unit (CPU). Within the CPU is a control unit that decodes the program into instructions, which are enacted by an arithmetic and logic unit (ALU), using data to perform calculations and tasks. The results of these are then fed back into the memory unit.

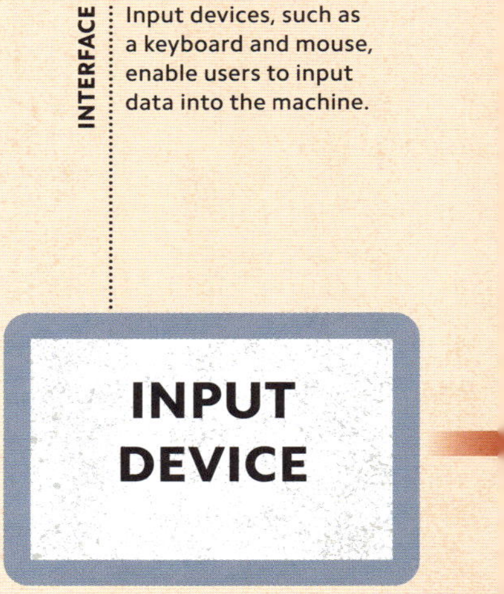

INTERFACE

Input devices, such as a keyboard and mouse, enable users to input data into the machine.

## INPUT DEVICE

**Structural advantage**
This diagram shows von Neumann's architecture. As the memory units could be upgraded, the machines could be made faster and more powerful.

> "Any computing machine that is to solve a complex mathematical problem must be 'programmed' for this task."
> John von Neumann

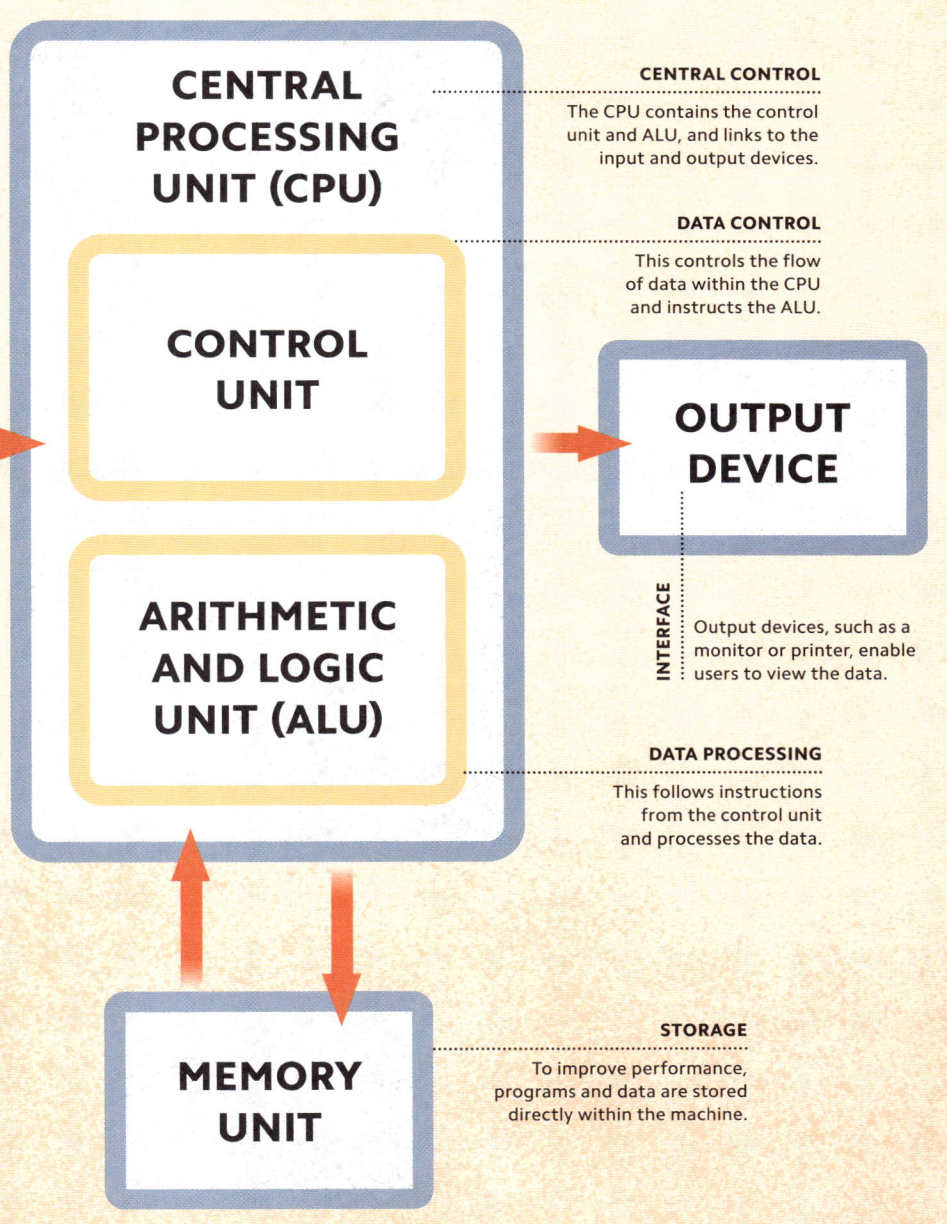

**CENTRAL PROCESSING UNIT (CPU)**

**CONTROL UNIT**

**ARITHMETIC AND LOGIC UNIT (ALU)**

**OUTPUT DEVICE**

**MEMORY UNIT**

**CENTRAL CONTROL**

The CPU contains the control unit and ALU, and links to the input and output devices.

**DATA CONTROL**

This controls the flow of data within the CPU and instructs the ALU.

**INTERFACE**

Output devices, such as a monitor or printer, enable users to view the data.

**DATA PROCESSING**

This follows instructions from the control unit and processes the data.

**STORAGE**

To improve performance, programs and data are stored directly within the machine.

# TWO KINDS OF AI

Whether the brain is a kind of living computer or not (see p.12), human intelligence and consciousness are the benchmarks that scientists use to measure AI capabilities. Some scientists argue that "weak" AI – which includes computers that can do specific, limited tasks, such as play chess or translate languages – is the only kind of AI that could ever be built. Others believe that, one day, "strong" AI – an intelligence that can match a human being's in every way – will be a reality. Such an AI would not only possess human-like cognitive abilities, it might, its defenders argue, be conscious (see pp.128–129) – and so could be accorded rights (see p.135).

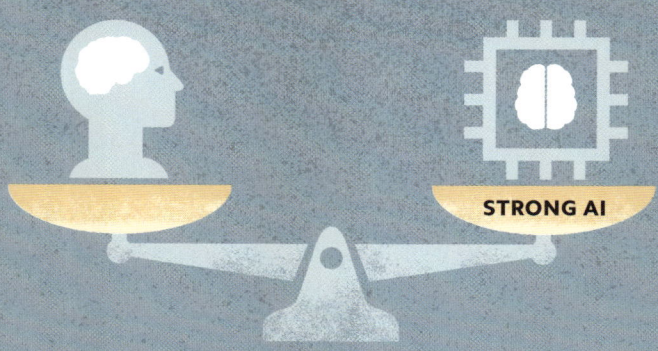

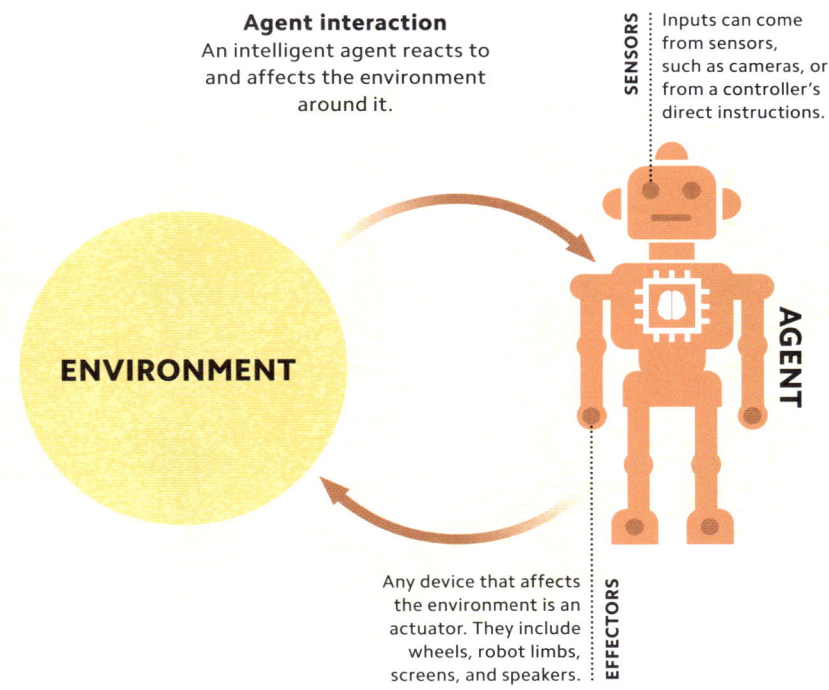

**Agent interaction**
An intelligent agent reacts to and affects the environment around it.

**SENSORS**
Inputs can come from sensors, such as cameras, or from a controller's direct instructions.

**ENVIRONMENT**

**AGENT**

Any device that affects the environment is an actuator. They include wheels, robot limbs, screens, and speakers.

**EFFECTORS**

# AI IN ACTION

An "intelligent agent" in AI is anything that can sense, respond to, and affect its environment – which can be physical or digital. Examples include robots, thermostats, and computer software programmes. The agent has "sensors", which it uses to perceive its environment, and "actuators", which it uses to interact with its surroundings. The action the agent takes depends on the specific goals it has been set and on what it senses. Some agents can learn (see pp.58–59), so that they are able to change the way they react to conditions within their environment.

# TRIAL AND ERROR

Machines that can follow simple instructions, such as calculators that apply mathematical rules, have existed for decades. Creating machines that can "learn" – the basis of modern AI – is far more recent and complex. To do so, programmers use algorithms (see p.14) that are repeatedly revised through trial and error to improve their accuracy. Like natural evolution, the improvements made are gradual and incremental. As AIs become more advanced, they are able to contribute to their own learning, although currently they require human assistance.

**Improved accuracy**
Teaching machines to learn means making them more accurate and more reliable.

# MIMICKING THE BRAIN

Connectionism is an approach to AI in which information is represented not by symbols but by patterns of connection and activity in a network. These patterns are known as "distributed representations", and computation that is done in this way is known as "parallel distributed processing" (PDP). Connectionists believe that intelligence can be achieved by taking simple processing units, such as artificial neurons (see p.21), and connecting them together into huge "artificial neural networks" (ANNs, see p.76) to allow PDP. As its name suggests, the connectionist model is based on how the brain works – using parallel processing across interconnected networks of cells, or neurons.

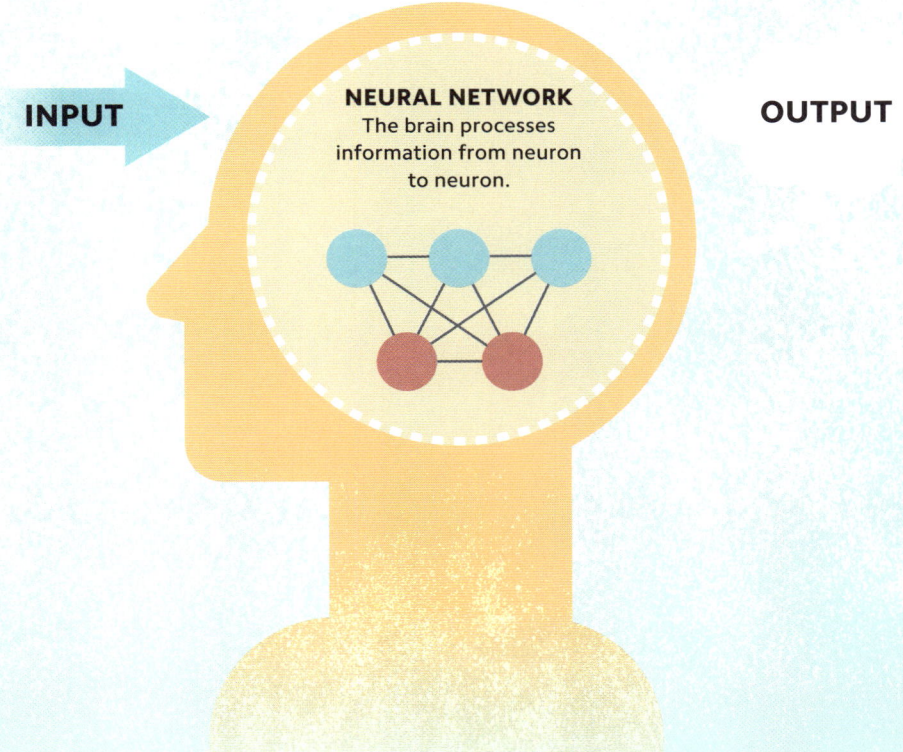

**INPUT**

**NEURAL NETWORK**
The brain processes information from neuron to neuron.

**OUTPUT**

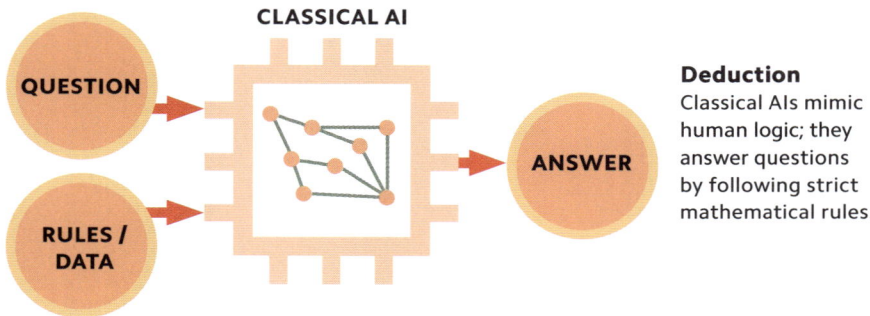

**CLASSICAL AI**

QUESTION

RULES / DATA

ANSWER

**Deduction**
Classical AIs mimic human logic; they answer questions by following strict mathematical rules.

# AI MODELS

The earliest forms of AI are now known as classical (or symbolic) AIs. They were constructed according to the top-down approach, in which computer designers first worked out the rules of symbolic reasoning – how humans think – and built them into the AIs. Their performance was always limited by the rigid application of human-derived rules, and their programmers' understanding of them. In contrast, modern statistical AIs are built according to the bottom-up approach. They are provided with masses of data and machine-learning tools (see pp.58–59) that enable them to find patterns in the data. From these patterns they are able to build models that show how particular systems (such as financial markets) operate under particular conditions.

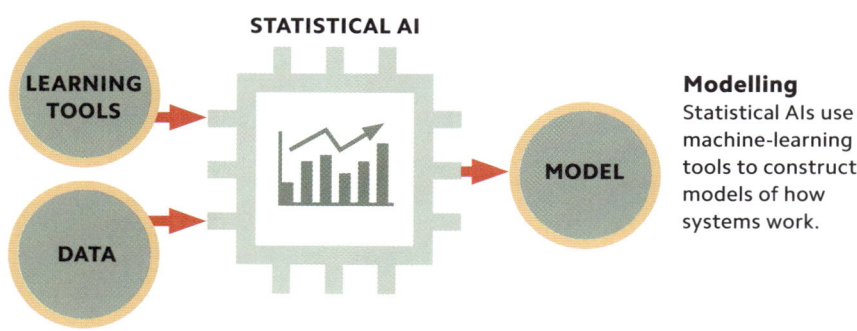

**STATISTICAL AI**

LEARNING TOOLS

DATA

MODEL

**Modelling**
Statistical AIs use machine-learning tools to construct models of how systems work.

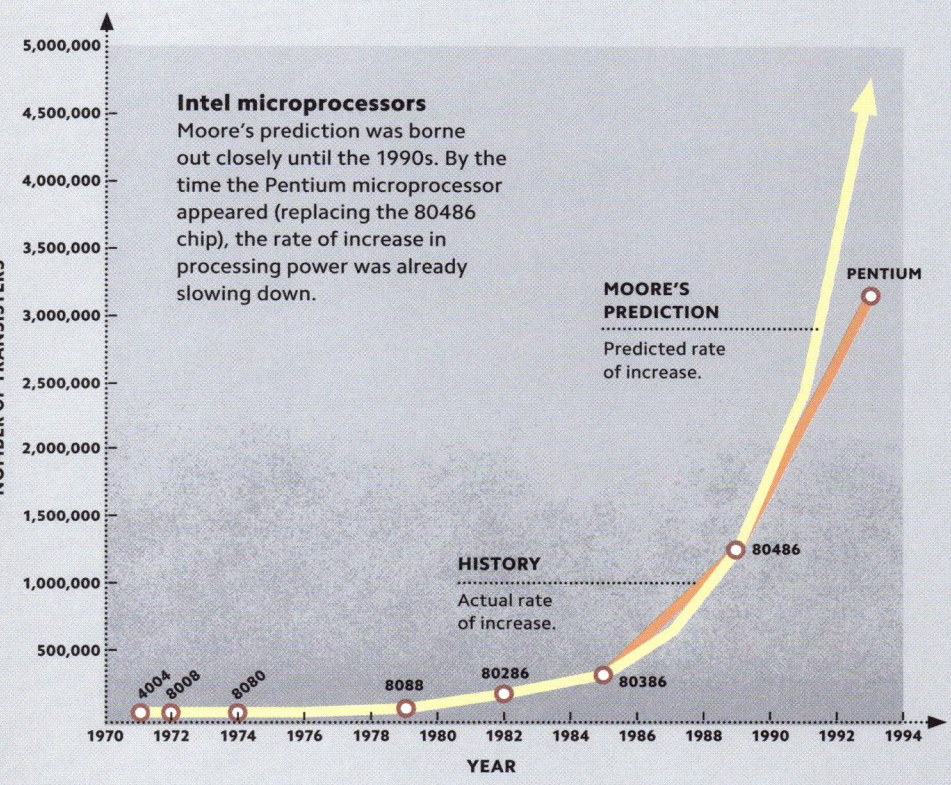

**Intel microprocessors**
Moore's prediction was borne out closely until the 1990s. By the time the Pentium microprocessor appeared (replacing the 80486 chip), the rate of increase in processing power was already slowing down.

**MOORE'S PREDICTION**
Predicted rate of increase.

PENTIUM

**HISTORY**
Actual rate of increase.

80486

4004
8008
8080
8088
80286
80386

NUMBER OF TRANSISTERS

5,000,000
4,500,000
4,000,000
3,500,000
3,000,000
2,500,000
2,000,000
1,500,000
1,000,000
500,000

1970 1972 1974 1976 1978 1980 1982 1984 1986 1988 1990 1992 1994
YEAR

# COMPUTING POWER

Moore's Law is named after Gordon Moore (1929–), the cofounder of integrated circuit chip-maker Intel. In 1965, Moore predicted that the number of transistors that could be fitted onto a computer chip would double every two years. Due to advances in technology, particularly miniaturization, this prediction was borne out for decades, and although it has since slowed down, computing power is still increasing each year. This means that in the foreseeable future, if computationalism is correct (see p.12), AIs will have the same amount of computing power as the human brain.

# RAW INFORMATION

Data is information that can take many forms, such as numbers, words, or images. In computing, data is a sequence of symbols that is collected and processed by a computer according to its programming. In modern computers, these symbols are the 1s and 0s of binary – or digital – code (see p.13). This data is either "at rest" (stored physically in a database), "in transit" (being used for a finite task), or "in use" (constantly being updated), and it can also be shared between computers. Data is classified according to whether it can be measured and how this is done.

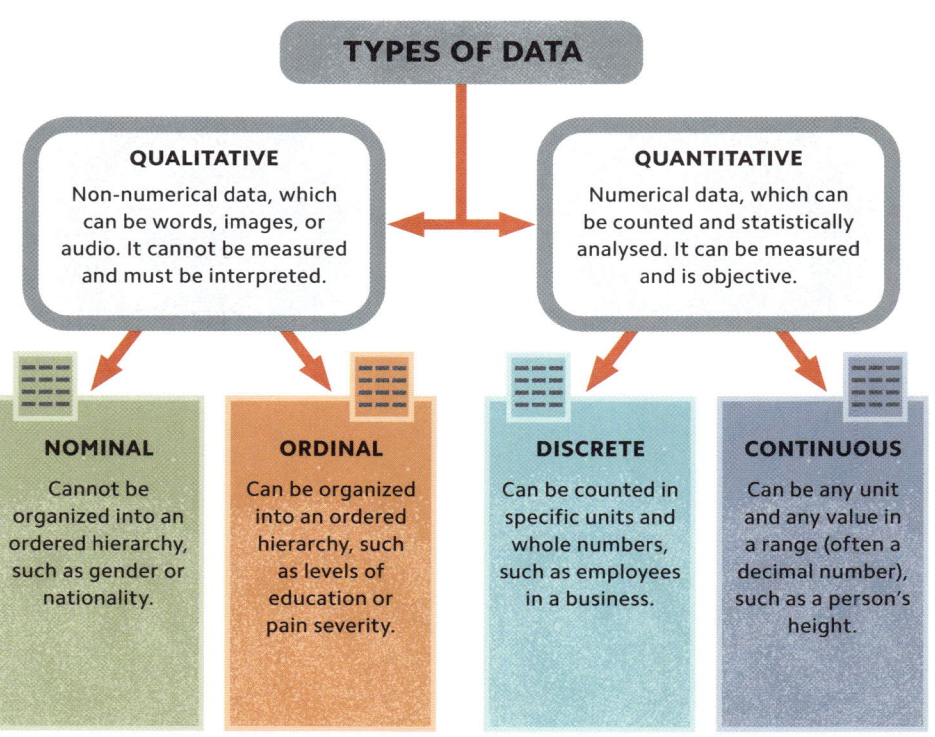

**TYPES OF DATA**

**QUALITATIVE**
Non-numerical data, which can be words, images, or audio. It cannot be measured and must be interpreted.

**QUANTITATIVE**
Numerical data, which can be counted and statistically analysed. It can be measured and is objective.

**NOMINAL**
Cannot be organized into an ordered hierarchy, such as gender or nationality.

**ORDINAL**
Can be organized into an ordered hierarchy, such as levels of education or pain severity.

**DISCRETE**
Can be counted in specific units and whole numbers, such as employees in a business.

**CONTINUOUS**
Can be any unit and any value in a range (often a decimal number), such as a person's height.

# EVERYTHING, EVERYWHERE, ALL OF THE TIME

"Big data" is a phrase that describes data sets that are too large to be processed by traditional forms of data-processing software. Such data sets include massive amounts of information about people, their behaviour, and their interactions. For example, mobile phone companies use their customers' phones to track the movements of billions of people, every second of every day, and they record this information in vast data sets. Big data is widely used in AI, from training machine-learning models (see pp.58–59), making predictions about the weather or future customer behaviour (see pp.70–71), to protecting against cyberattacks (see p.97).

CLASSIC
ARTIFIC
INTELLI

A L

I A L

G E N C E

**From the 1950s to the 1990s**, the dominant paradigm in AI research was classical (or "symbolic" or "logical") AI. This approach to AI was based on logical reasoning, using symbols and rules – written by human programmers – to represent concepts and the relationships between them. Classical AI had many successes, including AIs that could play games, hold basic conversations, and answer queries using "expert systems". Although statistical AI has since overtaken classical AI, the old approach has not been entirely abandoned; many of its techniques have been incorporated into modern AI applications, such as natural language processing and robotics.

# REPRESENTING DATA

In AI, a "symbol" is a graphical representation of a real-world item or concept – a simple type of symbol is a picture. A symbol can also be a group of other symbols, such as the letters that make up the name of an object. In classical AI, symbols embody the total sum of the relevant facts and information required for the system to understand what something is. To achieve this, data is labelled (see pp.62–63) and attached to a symbol. The symbol for an apple would include a wealth of data stating what an apple is and is not.

# FOLLOWING THE RULES

Logic is the study of sound reasoning, and of the rules that determine what makes an argument valid. In practice, logic enables people to take statements about the world (known as premises) and derive new information from those statements (known as conclusions). AIs are programmed to follow strictly logical rules, with the aim of producing reliable conclusions. One such rule is the syllogism, which states: "If all As are Bs and all Bs are Cs, then all As are Cs". This simple principle enables AIs to know that all items of a particular class will always have a particular characteristic.

### Syllogistic logic
An AI that understands that fruit is healthy, and that an apple is a fruit, also knows that apples are healthy.

**PREMISE 1:**
**APPLES ARE FRUIT**

**CONCLUSION**
**APPLES ARE HEALTHY**

**PREMISE 2:**
**FRUIT IS HEALTHY**

# WHAT, WHEN, WHY, AND HOW?

AI systems use up to five kinds of knowledge in their interactions with the world, but only two are common to all AIs. Declarative knowledge is the most basic form and describes statements of fact, such as "cats are mammals", whereas procedural knowledge instructs AIs how to complete specific tasks. In some AIs, meta-, heuristic (see p.43), and structural knowledge provide further information that enables them to solve problems.

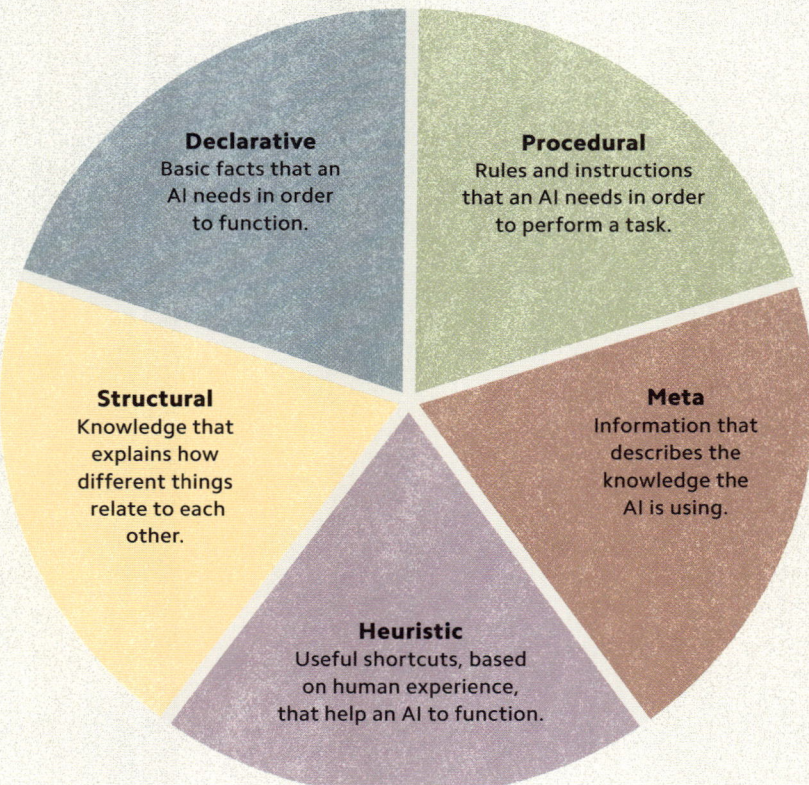

**Declarative**
Basic facts that an AI needs in order to function.

**Procedural**
Rules and instructions that an AI needs in order to perform a task.

**Structural**
Knowledge that explains how different things relate to each other.

**Meta**
Information that describes the knowledge the AI is using.

**Heuristic**
Useful shortcuts, based on human experience, that help an AI to function.

# PRESENTING KNOWLEDGE

In order for an AI to understand information correctly, the information must be presented to it very clearly. There are four main ways of doing this. "Logical representation" poses information using the exact words of a natural language (or symbols to represent them). "Semantic representation" ensures that the individual meanings within the information are connected in a formal, logical way. "Frame representation" involves presenting the information in a tabular format, with facts allocated to individual "slots". Finally, "production rules" are the instructions that state what conclusions an AI can deduce from the information it is supplied with (see p.37).

### Logical representation
Statements of information are clear, logical, and unambiguous.

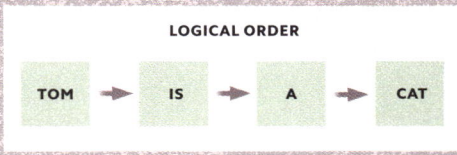

### Semantic representation
The relationships and connections between facts within the information are made clear.

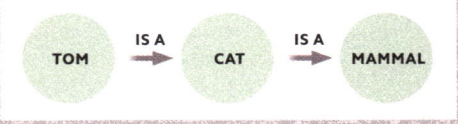

### Frame representation
Information can be presented as simple tables. Here, the slots contain details about Tom the cat.

### Production rules
When an "IF" statement is true, a "THEN" statement can be deduced from it.

| IF | THEN |
| --- | --- |
| TOM IS A CAT | TOM IS A MAMMAL |

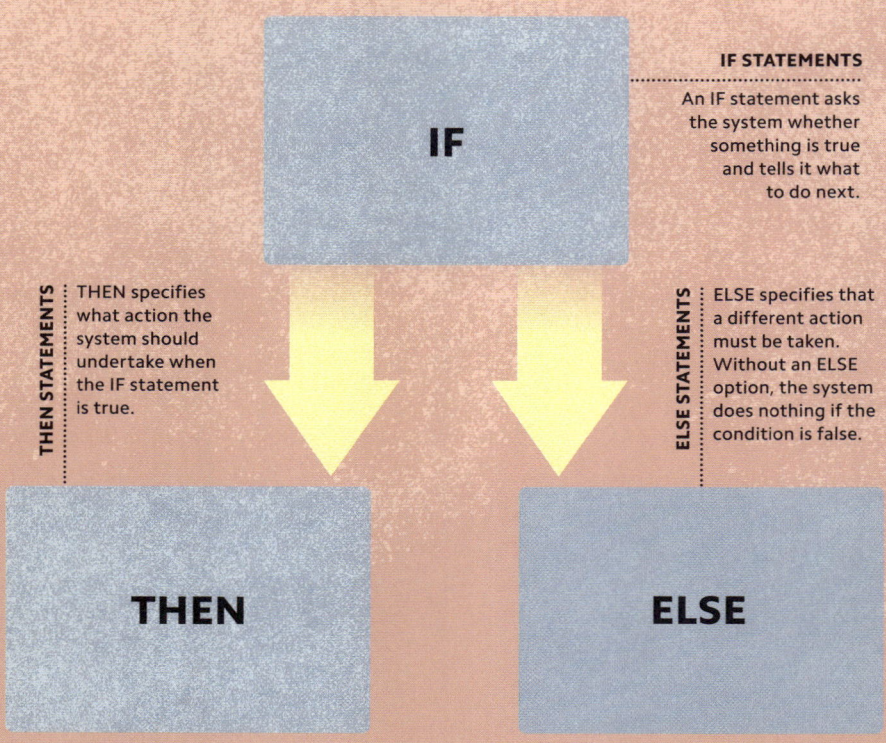

IF

IF STATEMENTS

An IF statement asks the system whether something is true and tells it what to do next.

THEN STATEMENTS

THEN specifies what action the system should undertake when the IF statement is true.

ELSE STATEMENTS

ELSE specifies that a different action must be taken. Without an ELSE option, the system does nothing if the condition is false.

THEN

ELSE

# IF THIS, THEN THAT

A rule-based AI system uses instructions, consisting of "IF-THEN" statements, to draw conclusions based on an initial set of facts. In its simplest form, an IF-THEN statement says to the system: "if this condition is true for the current facts, then do this; if it is false, do nothing". Adding an "ELSE" option allows for more complicated statements: "if this is true, then do this; otherwise (else), do that". Rule-based systems are predictable, reliable, and "transparent", meaning it is easy to see which rules the AI applies. However, rule-based AIs cannot "learn" by adding to their store of rules and facts without human intervention.

> "Much of what we do with machine learning happens beneath the surface."
> Jeff Bezos

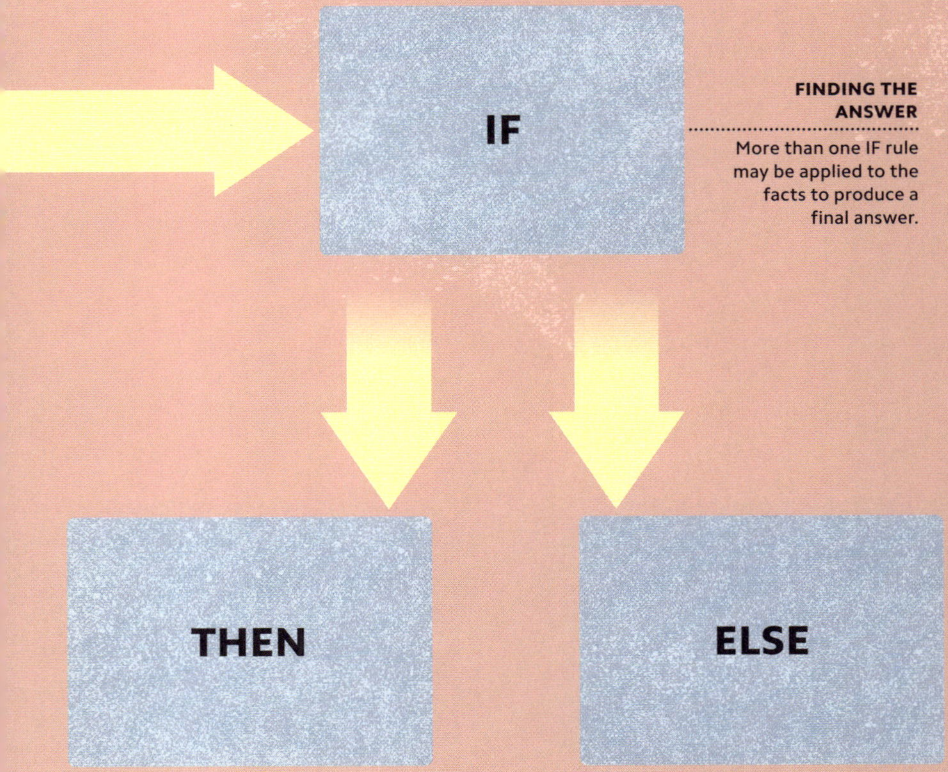

**IF**

**THEN**

**ELSE**

**FINDING THE ANSWER**

More than one IF rule may be applied to the facts to produce a final answer.

# THE SHORTEST ROUTE

Pathfinding algorithms are search algorithms that are used to find the shortest route between two points. They have many uses, including in vehicle navigation and computer gaming. The algorithm is programmed using a weighted graph (see below) that shows all of the possible paths available. The circles, or "nodes", represent waypoints, or special locations, which are joined by lines known as "edges". Programmers add a weight (see p.78) to the edges, which reflects a "cost", such as distance or time. The algorithm calculates the weights to find the shortest path.

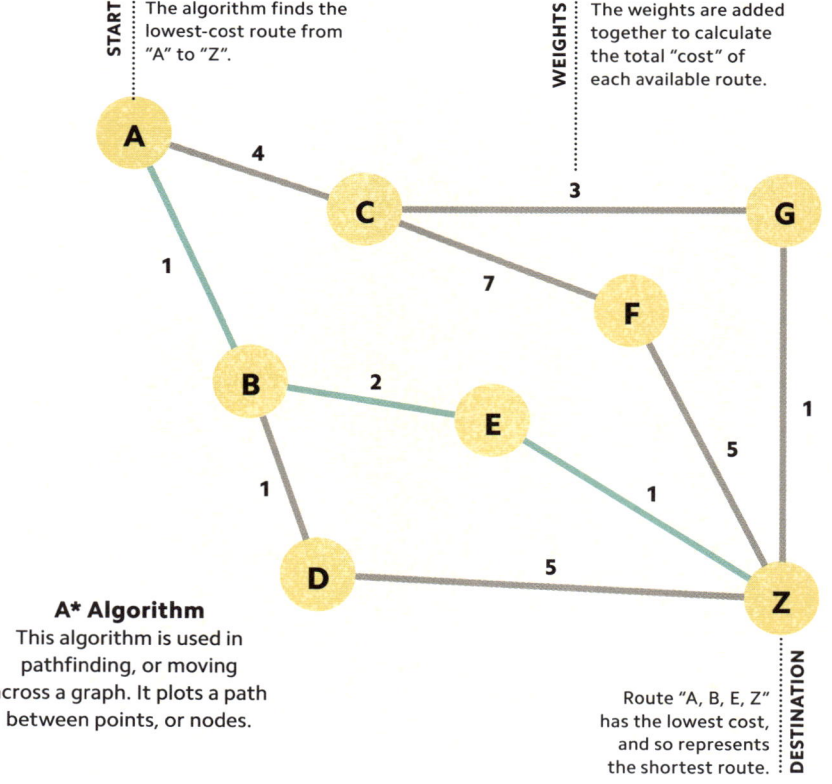

**START** The algorithm finds the lowest-cost route from "A" to "Z".

**WEIGHTS** The weights are added together to calculate the total "cost" of each available route.

**A\* Algorithm**
This algorithm is used in pathfinding, or moving across a graph. It plots a path between points, or nodes.

Route "A, B, E, Z" has the lowest cost, and so represents the shortest route.

**DESTINATION**

# IMPERFECT SOLUTIONS

Some problems are too complex for an algorithm to solve quickly. In such cases, an AI can do a "brute-force search", which means to methodically work through and evaluate every possible solution. This is slow, however, and in some cases impossible. A more efficient alternative is to use a "heuristic". This practical method uses a common-sense approach, searching for an approximate solution by estimating a "good enough" choice at every decision point based on the information available.

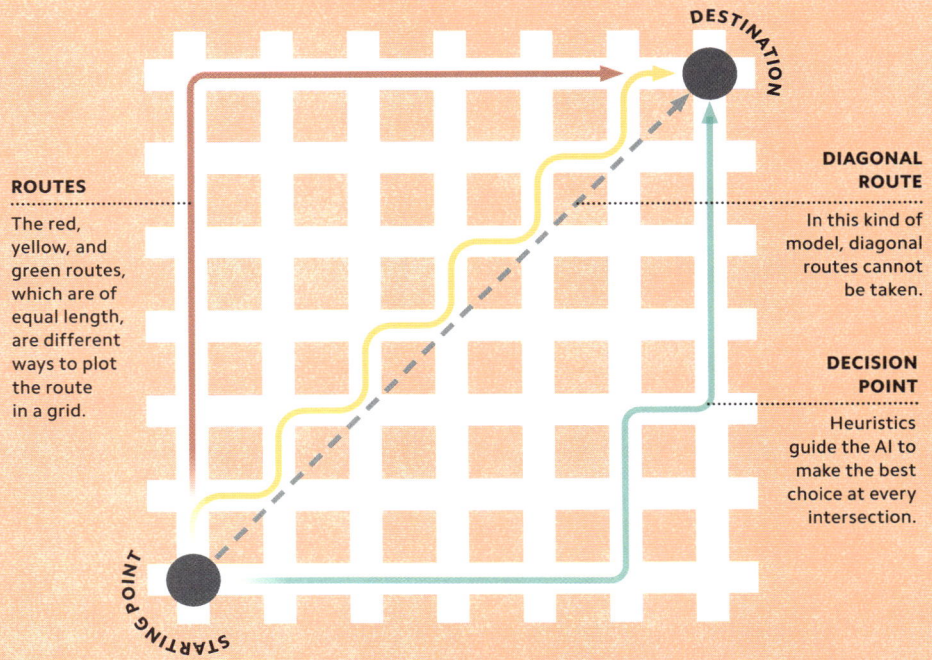

**ROUTES**

The red, yellow, and green routes, which are of equal length, are different ways to plot the route in a grid.

DESTINATION

**DIAGONAL ROUTE**

In this kind of model, diagonal routes cannot be taken.

**DECISION POINT**

Heuristics guide the AI to make the best choice at every intersection.

STARTING POINT

**Manhattan distance**
The Manhattan distance heuristic maps routes by calculating squares moved vertically and horizontally. It can be used to plot a path in a city with a grid system, such as Manhattan, New York.

# PERFORMING A TASK

Embodied AIs (see p.118), such as robots, use a technique known as "planning" to help them solve practical problems. Planning involves understanding the environment or location in which the task must be performed and mapping out the actions required to complete it. The AI must identify each step required to fulfil the task and the optimal (lowest-cost) sequence in which to perform them (see p.42). If the optimal sequence is not possible, for whatever reason, it must be able to decide the next-best alternative (see p.43). It must also identify and avoid any actions that would prevent it from completing its task.

**Planning ahead**
In order to complete its task, the robot breaks it down into a sequence of individual steps.

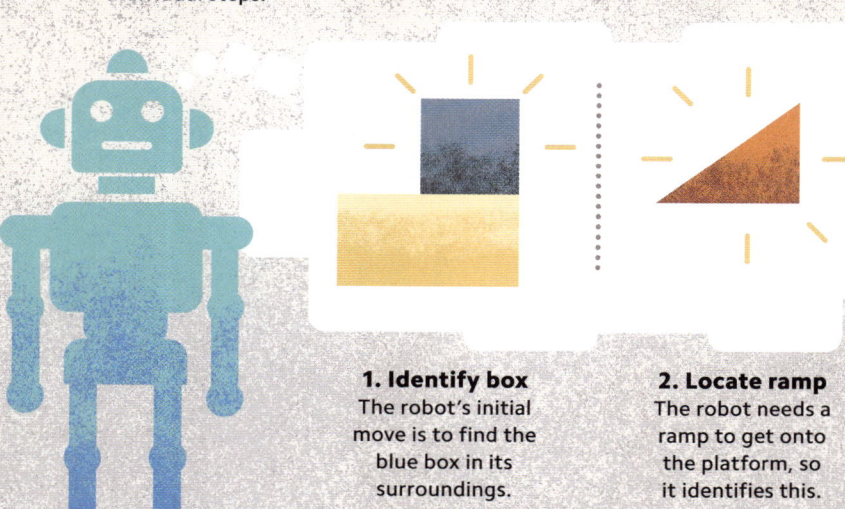

**1. Identify box**
The robot's initial move is to find the blue box in its surroundings.

**2. Locate ramp**
The robot needs a ramp to get onto the platform, so it identifies this.

### ENVIRONMENT
The robot must climb onto the platform to reach the blue box.

### Task
The robot's goal is to push the blue box off the end of the platform.

### 3. Push ramp
The ramp needs to be adjacent to the platform, so it must be pushed into place.

### 4. Ascend ramp
The robot can now use the ramp to move up onto the platform next to the blue box.

### 5. Push block off
Once on the platform, the robot can push the box off the end. Its task is now complete.

**Bayes' theorem**
The probability of one event happening – such as smoke accompanying a dangerous fire – depends on previous events, including the known frequency of smoke and fires.

FIRE WITH SMOKE

The probability of event A happening given that event B has happened. For example, the probability that a fire is dangerous, given that smoke is present.

$$P(A|B) =$$

# DEALING WITH UNCERTAINTY

Most classical AIs are based on the idea that logical statements (see p.37) are either true or false – that there is no room for uncertainty. However, uncertainty is an unavoidable feature of life, and it can be incorporated into AIs using the concept of probability. Probability is a numerical value of how likely something is to occur. "Probabilistic reasoning" is any method of reasoning that takes probability into account. The English statistician Thomas Bayes (1702–61) developed a method, known today as Bayes' theorem, of calculating the likelihood of an event happening. Rather than working out the probability of the event in isolation, Bayes' theorem bases probability on prior knowledge of the relevant conditions.

The probability of event B happening given event A has happened. For example, the likelihood that there is smoke accompanying a dangerous fire.

Probability of event A occurring. For example, how often dangerous fires occur.

$$\frac{P(B|A)\,P(A)}{P(B)}$$

Probability of event B happening. For example, how often there is smoke.

"Probability theory is nothing more than common sense reduced to calculation."
Pierre-Simon Laplace

# MODELLING CHANGES

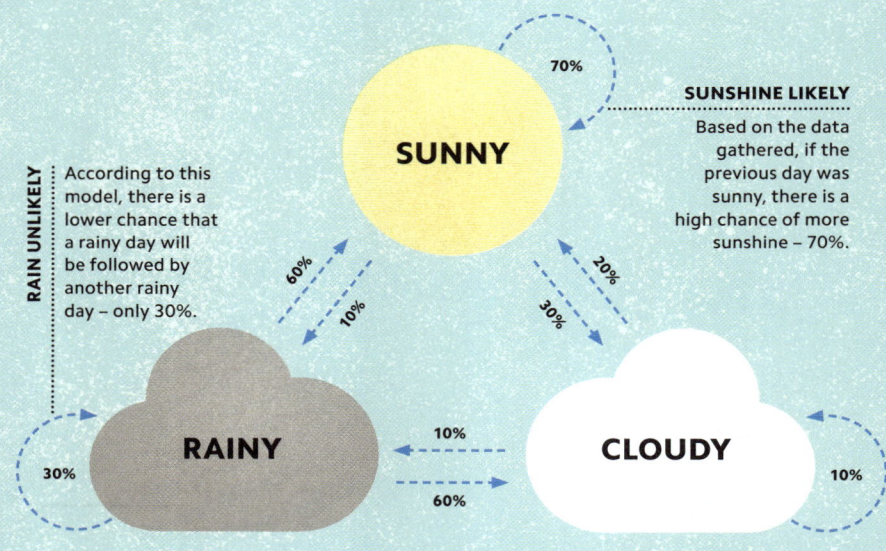

70%

**SUNSHINE LIKELY**
Based on the data gathered, if the previous day was sunny, there is a high chance of more sunshine – 70%.

**SUNNY**

**RAIN UNLIKELY**
According to this model, there is a lower chance that a rainy day will be followed by another rainy day – only 30%.

60%

10%

20%

30%

**RAINY**

30%

10%

60%

**CLOUDY**

10%

A Markov chain is a model that describes a sequence of possible events in which the probability of each event depends on the state that was reached in the previous event. The model predicts outcomes based on the rules of probability (see pp.46–47) and using data collected on the relevant subject. Once it has been trained (see p.61), it only needs to know the conditions of the immediate past (the previous state) to get the relevant information to predict the likelihood of the next state. Markov chains have many AI applications, from forecasting weather patterns and financial market conditions, to use in predictive text systems.

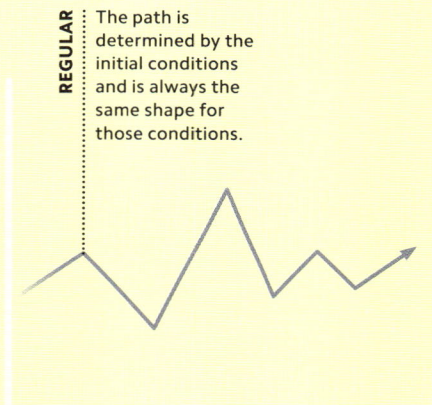

The path is determined by the initial conditions and is always the same shape for those conditions.

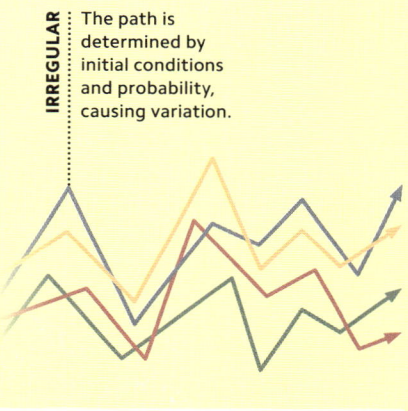

The path is determined by initial conditions and probability, causing variation.

**Deterministic models**
In a deterministic model, there are no random variables. Results from a set of inputs will be related in a predictable way.

**Stochastic models**
A stochastic model includes random variables. Results are much less predictable and not clearly related to each other.

# MODELLING UNCERTAINTY

Stochastic models enable AIs to make predictions about processes and situations that are affected by chance events, such as changes in the stock market or the growth rate of bacteria. The volatile and ever-changing factors in these scenarios are represented by random variables, and each is assigned a value based on the probability of it occurring. A stochastic model then processes thousands of combinations of variables and produces a distribution curve that shows the probability of different outcomes under different circumstances.

# AUTOMATED ADVICE

Computer programs that replicate the knowledge and reasoning skills of human specialists are known as "expert systems". The information that they contain is supplied by human experts, and is programmed into the system by a "knowledge engineer". Each system has three parts. The "knowledge base" contains the facts and rules used by human experts on the topic. The "inference engine" applies the rules to the facts in the knowledge base to deduce answers to queries posed by users. The "user interface" accepts queries from users and displays the solutions found by the system. Expert systems are able to answer complex questions and provide users with wider access to expert advice. They are used in many areas, including medicine, where they match symptoms to likely causes and appropriate treatments.

**BUILDING PHASE**

**Human experts**
Experts supply the knowledge and rules within the system.

**Knowledge engineer**
The expert system is programmed by a knowledge engineer.

**OPERATING STATE**

**User**
The user asks a question and gets an answer via the interface.

QUERY

ANSWER

**Inference engine**
An inference engine applies rules to facts in the knowledge base, matching a user's question to potential answers.

**User interface**
The user interface is the software that the user interacts with. For example, the user can describe symptoms and then receive a diagnosis.

**Knowledge base**
A knowledge base is an organized collection of facts about a particular subject, such as medicine.

**In action**
The three sections of an expert system interact to provide answers to the user.

# HANDLING "MESSY" DATA

Classical AIs (see p.30) struggle with some tasks that humans find simple. We can programme computers for reasoning-based tasks, such as playing chess, but not for sensorimotor- and perception-based tasks such as catching a ball or recognizing a cat.

The Austrian–Canadian programmer Hans Moravec (1948–) argued that reasoning tasks are easy to teach to computers because humans have already worked out the steps that are required to complete them. In contrast, sensorimotor and perception activities involve unstructured, or "messy", data that requires a lot of processing. For humans, these tasks are largely unconscious actions, refined over millions of years of brain evolution, but they are difficult to break down into a series of steps that a computer can follow.

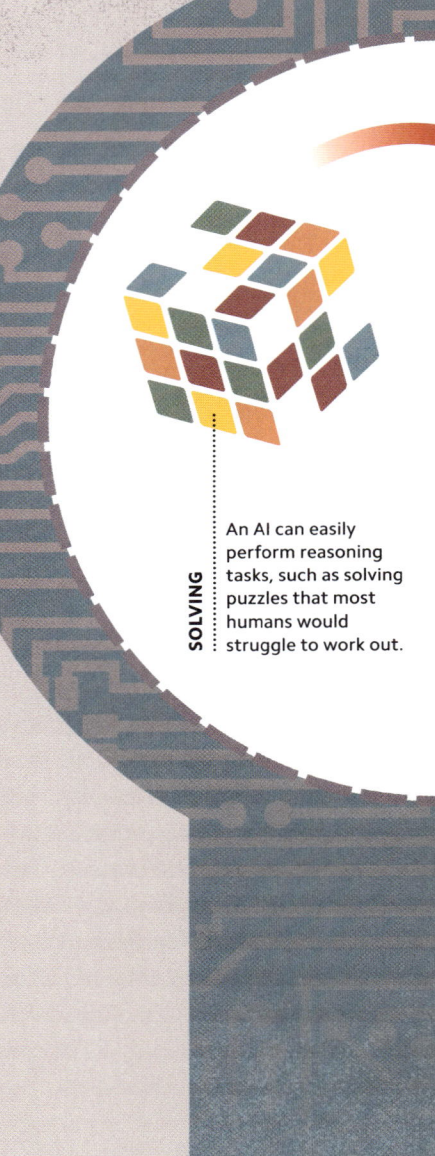

SOLVING

An AI can easily perform reasoning tasks, such as solving puzzles that most humans would struggle to work out.

A human can easily manipulate a physical object, but this is a complex challenge for AIs, particularly classical models.

# NEATS VS

In the 1970s, AI theorist Roger Schank (1946–) noted that there are two types of AI research, which he called "neat" and "scruffy" (see opposite). The neat approach, which has since become dominant, builds AIs by programming computers to follow strict mathematical rules. These rules enable AIs to distinguish between different types of data, and to analyse those data by using machine-learning algorithms (see pp.58–59). Artificial neural networks (ANNs, see p.76), for example, are a triumph of the neat approach.

**Neat AI**
Defenders of the neat approach argue that AIs are machines that can perform specific tasks with complete reliability. They also claim that neat AIs will ultimately have human-like intelligence.

**PREDICTABLE**
.........................................
Neat designers take their cue from physics, building AIs whose behaviour is predictable.

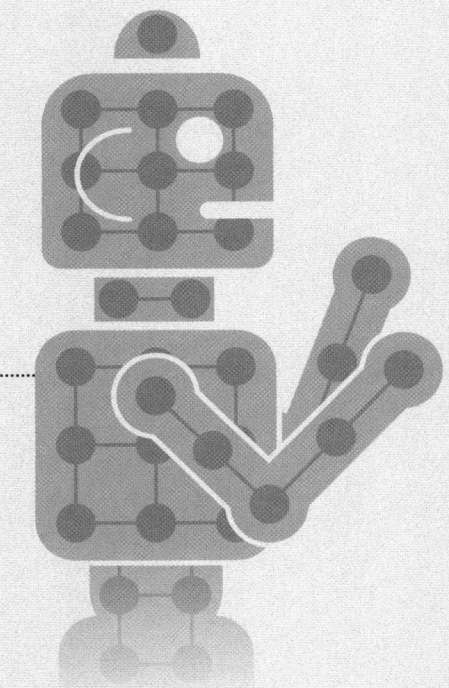

# SCRUFFIES

Roger Schank (see opposite) defined the "scruffy" approach to AI as a method in which researchers experiment with all kinds of models and algorithms in order to design programs that show intelligence. Marvin Minsky (1927–2016) described this approach as being "analogical" rather than logical, since it embraces the idea that an AI, like a human being, should be able to recognize that certain problems are analogous (or comparable) to other problems – in other words, it should have a kind of common sense.

**Scruffy AI**
Defenders of the scruffy approach argue that a scruffy AI is more likely to achieve human-like intelligence than a neat AI, because there is more to human intelligence than following rules.

**LESS PREDICTABLE**
Scruffy designers take their cue from biology, building AIs that are less predictable than neats.

STATIST

ARTIFIC

INTELLI

# ICAL
# IAL
# GENCE

**In the 1990s** many researchers grew frustrated by the shortcomings of classical AI, with its focus on logic and deductive reasoning, and began developing statistical techniques instead. This gave rise to statistical AI, which remains the main focus of AI research today. At the heart of this approach is a technique known as "machine learning". Machine learning involves using data sets to train AI models (including models that mimic the human brain, known as "artificial neural networks") to perform tasks without requiring an engineer to program them explicitly to do so. This approach is thriving today, due to the availability of powerful computer hardware and large data sets.

## ARTIFICIAL INTELLIGENCE

This is the science of developing machines that can act and make decisions "intelligently".

## MACHINE LEARNING

Machine learning focuses on training computers to perform tasks without the need for explicit programming.

## DEEP LEARNING

Deep learning is the most sophisticated type of machine learning. It requires minimal human intervention, and uses computer models known as "artificial neural networks" that are based on the human brain.

> **"Predicting the future isn't magic, it's artificial intelligence."**
> Dave Waters

# TEACHING AIs TO THINK

Machine learning is a form of AI that enables computer systems to learn how to perform tasks without being explicitly programmed to do so. Programmers can write algorithms that tell computers precisely which steps to follow to complete simple tasks. However, for more complex tasks, such as recognizing faces or understanding spoken conversations, it is incredibly difficult for programmers to write the necessary algorithms, and this is where machine learning comes in. Machine-learning algorithms use collections of sample data, known as training data (see p.61), to build models that make predictions or choices based on new data. There are many kinds of machine learning, including deep learning (p.86), in which AIs mimic the structure and behaviour of biological brains by using artificial neural networks (see p.76).

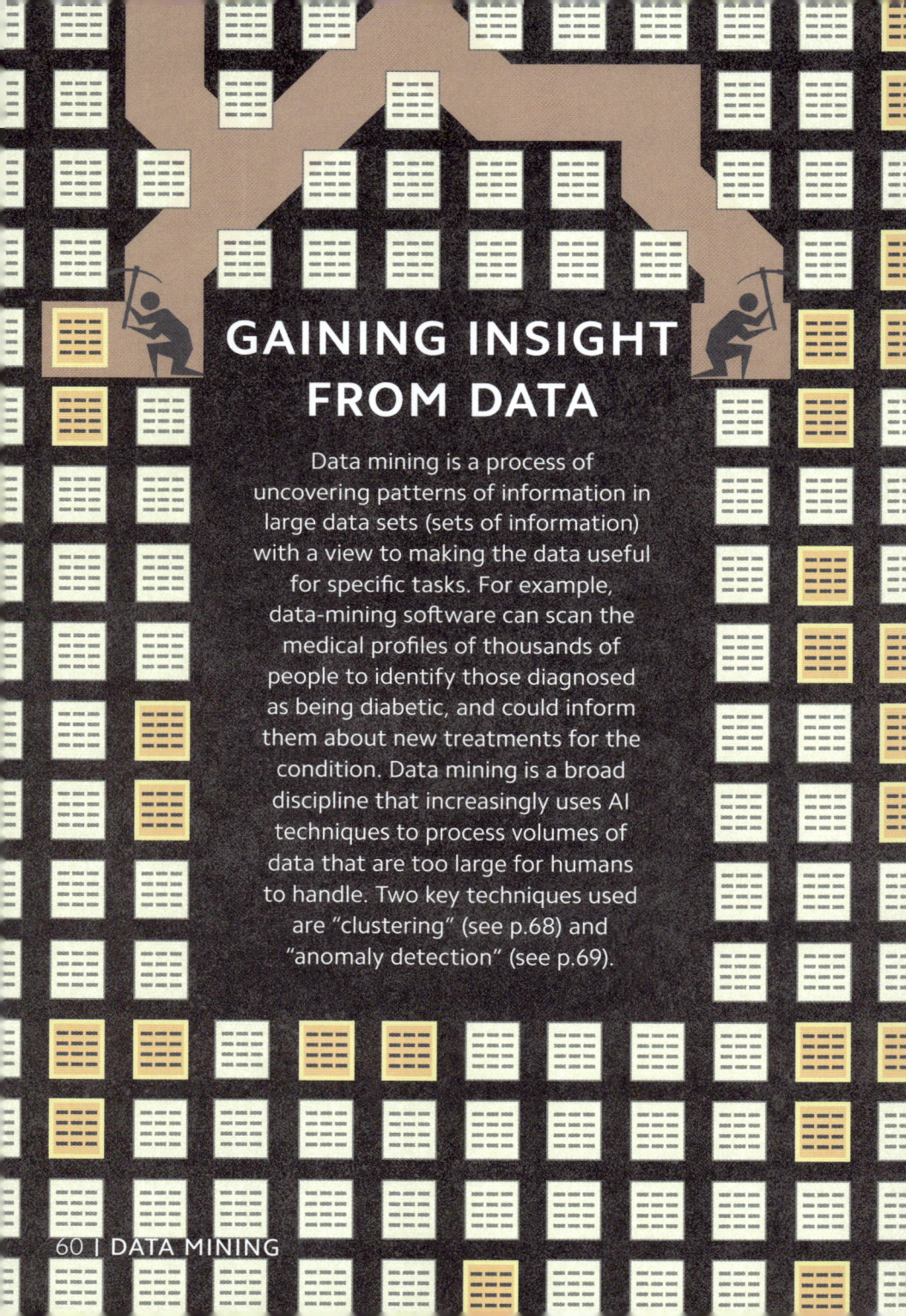

# GAINING INSIGHT FROM DATA

Data mining is a process of uncovering patterns of information in large data sets (sets of information) with a view to making the data useful for specific tasks. For example, data-mining software can scan the medical profiles of thousands of people to identify those diagnosed as being diabetic, and could inform them about new treatments for the condition. Data mining is a broad discipline that increasingly uses AI techniques to process volumes of data that are too large for humans to handle. Two key techniques used are "clustering" (see p.68) and "anomaly detection" (see p.69).

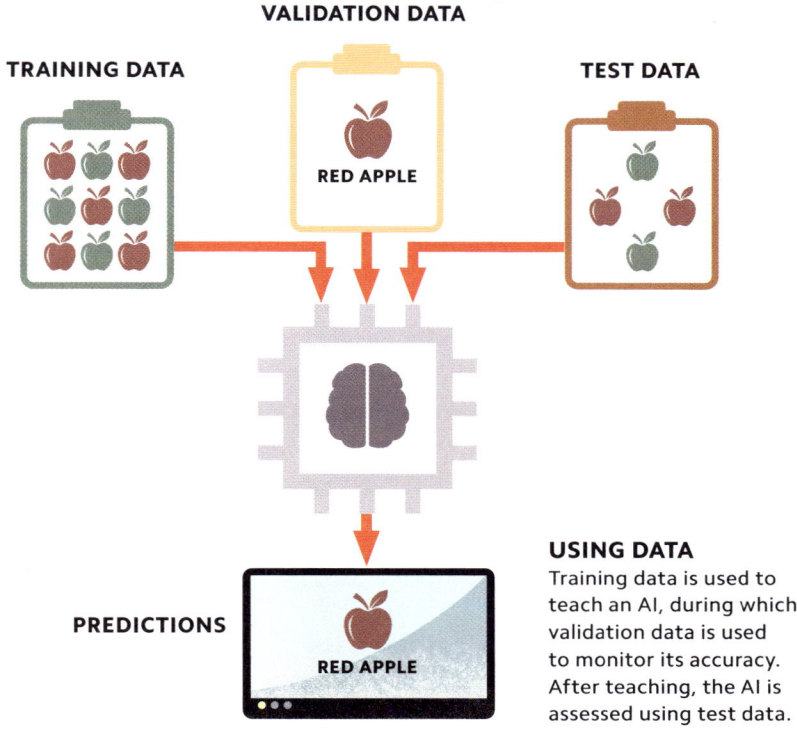

**TRAINING DATA**

**VALIDATION DATA**

**TEST DATA**

RED APPLE

**USING DATA**
Training data is used to teach an AI, during which validation data is used to monitor its accuracy. After teaching, the AI is assessed using test data.

**PREDICTIONS**

RED APPLE

# TEACHING MATERIALS

Training data is a type of data that is used during machine learning (see pp.58–59) to teach AIs how to perform tasks accurately. It is used by programmers to test, adjust, and fine-tune the AI (see pp.78–79) until it gives the expected results – or outputs. "Validation data" may also be used to assess how accurately the AI processes the training data during the learning period. Once the AI has been trained, "test data" is then used to assess the accuracy of its results. Machine learning requires a large amount of training data, which may be labelled or unlabelled (see pp.62–63).

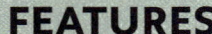

# FEATURES

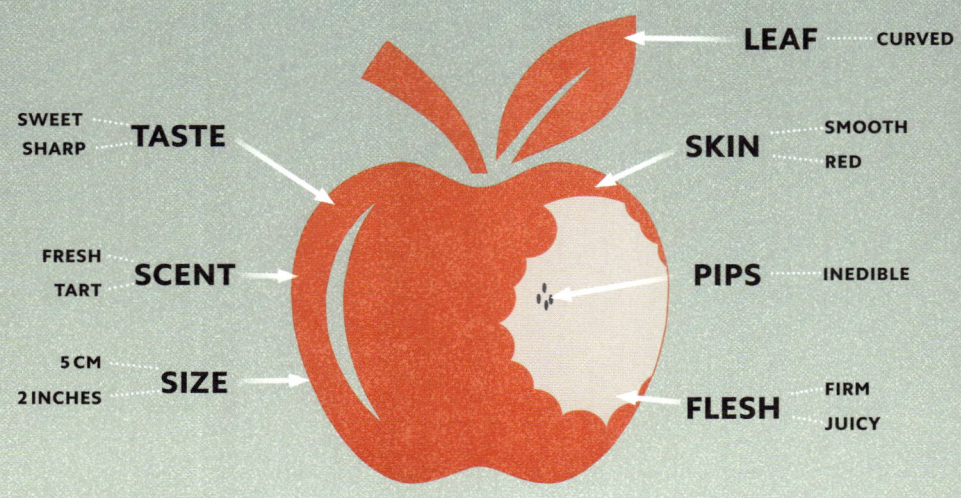

LEAF ........ CURVED

SWEET
SHARP **TASTE**

**SKIN**
SMOOTH
RED

FRESH
TART **SCENT**

**PIPS** ........ INEDIBLE

5 CM
2 INCHES **SIZE**

**FLESH**
FIRM
JUICY

**Tagging features**
A human operator tags all of the data
describing the features of "type A"
apples. The AI learns that, together,
these features define a "type A" apple.

# GIVING DATA MEANING

A "feature" is a characteristic, such as a pattern of pixels, that
an AI can use as an input to predict a label, which becomes the
output. In supervised machine learning (see p.72), AIs learn to
associate particular features with labels by processing training
data sets (see p.61) that have already been labelled by a human
operator. For example, if an image-recognition AI trained with
labelled photographs of animals is input a photograph of an
animal with features such as white feathers, curved beak,
and crest, it will probably output a label of "cockatoo".

**LABELS**

**FRUIT DATABASE SEARCH: APPLE A**

**Apple A**
Taste: sweet, sharp
Scent: fresh, tart
Size: 5 cm/2 inches
Leaf: curved
Skin: smooth, red
Pips: inedible
Flesh: firm, juicy

**Predicting labels**
Knowing all of the features of a "type A" apple, the AI can find it in a fruit database, and identify it with the label "Apple A".

"A baby learns to crawl, walk and then run. We are in the crawling stage when it comes to applying machine learning."
Dave Waters

# LOOKING FOR PATTERNS

Pattern recognition, which enables AIs to find answers within huge quantities of data, is one of the most versatile tools in machine learning (see pp.58–59). The AI is programmed to identify specified patterns or broader similarities within the data, which it can do far more quickly than a human. Pattern recognition can involve finding and sorting data into defined classes (see p.66) or grouping similar data more loosely into "clusters" (see p.68). It can also be used to identify how changing the inputs within an AI model will affect outputs, which is known as "regression" (see p.67).

# YES OR NO?

A decision tree is a model of the decision-making process used by AIs. It works by questioning data, to which the answers can only be "yes" or "no". One kind of decision tree is the "classification tree". By repeatedly posing yes/no questions, the AI splits the "root" (data set) into ever-smaller "branches" (subsets) that share particular features, until a single "leaf" (conclusion) is reached, pinpointing a specific classification within the data. Decision trees are commonly used in both machine learning (see pp.58–59) and data mining (see p.60).

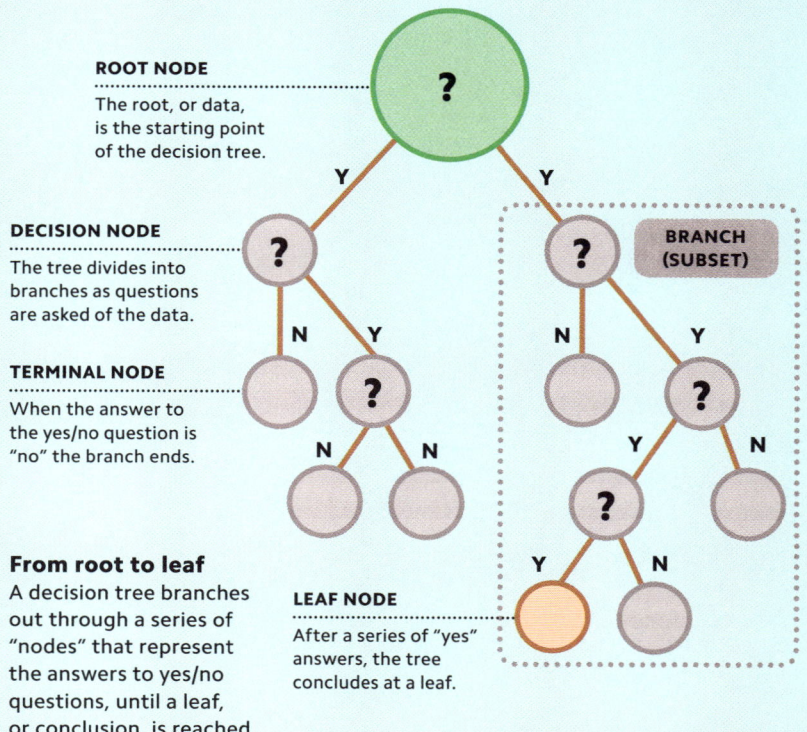

**ROOT NODE**
The root, or data, is the starting point of the decision tree.

**DECISION NODE**
The tree divides into branches as questions are asked of the data.

**TERMINAL NODE**
When the answer to the yes/no question is "no" the branch ends.

**BRANCH (SUBSET)**

### From root to leaf
A decision tree branches out through a series of "nodes" that represent the answers to yes/no questions, until a leaf, or conclusion, is reached.

**LEAF NODE**
After a series of "yes" answers, the tree concludes at a leaf.

# TYPES OF DATA

An algorithm that assigns labels to items (see pp.62–63) and then sorts them into categories, or "classes", is known as a "classifier". Through a process of supervised learning (see p.72), AIs are taught to classify items using a labelled training data set (see p.61), from which they learn to recognize the patterns associated with different labels. For example, a spam filter is taught to detect features of spam and non-spam emails from a collection of labelled emails. Based on this training data, the AI can automatically assign the labels "spam" or "not spam" to new emails.

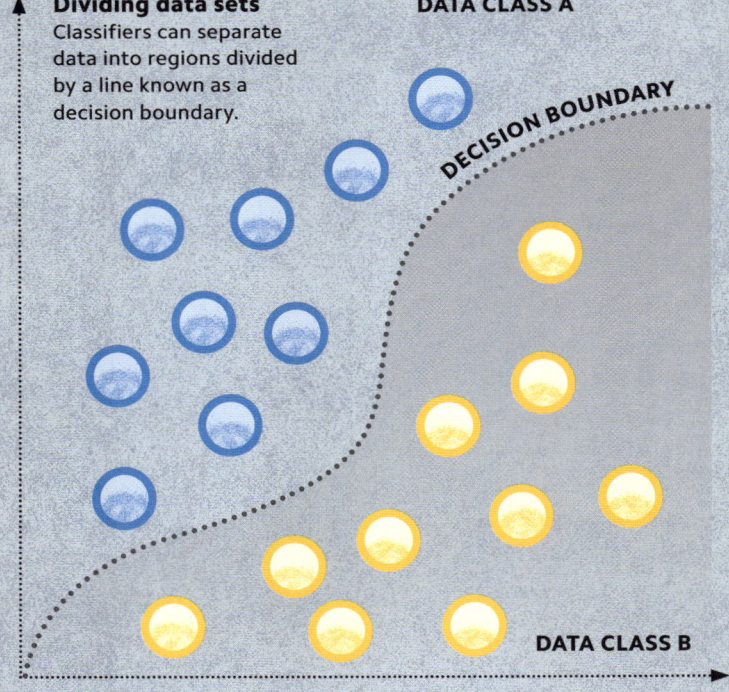

**Dividing data sets**
Classifiers can separate data into regions divided by a line known as a decision boundary.

DATA CLASS A

DECISION BOUNDARY

DATA CLASS B

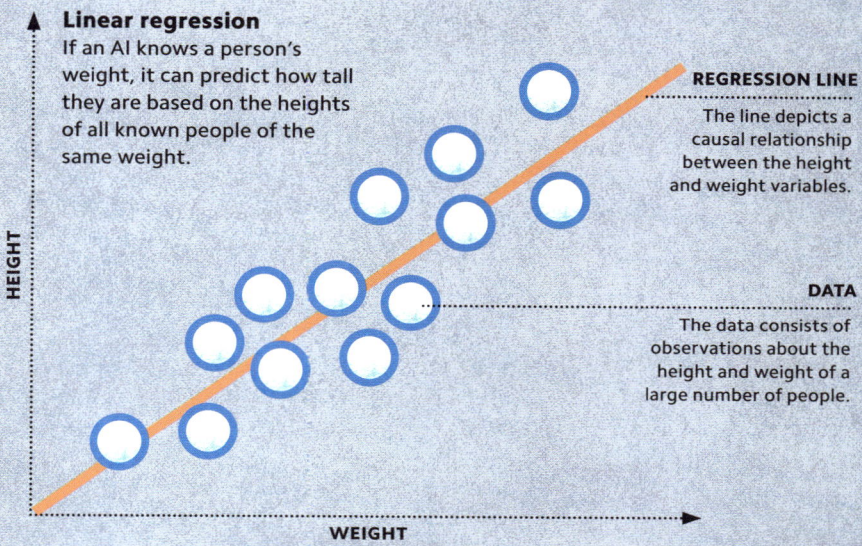

**Linear regression**
If an AI knows a person's weight, it can predict how tall they are based on the heights of all known people of the same weight.

HEIGHT

WEIGHT

**REGRESSION LINE**
The line depicts a causal relationship between the height and weight variables.

**DATA**
The data consists of observations about the height and weight of a large number of people.

# THE LINE OF BEST FIT

Regression is a process that is used in many fields, including machine learning (see pp.58–59), in which an algorithm is used to predict the behaviour of one or more variables depending on the value of another variable. It is used in many supervised learning applications (see p.72), particularly those that are designed to find causal relationships between several variables. For example, it can be used to predict what the next day's temperature will be given today's humidity, wind speed, and atmospheric pressure, and data about how all four variables have behaved in the past. "Linear regression" (see above) is the most common form of regression analysis, and is used particularly in the fields of finance and economics.

# GROUPING DATA

Clustering is the process of dividing a data set into a number of groups based on commonly shared features. It is an unsupervised machine-learning technique (see p.73), which means that it is performed by AIs on raw, unlabelled training data sets. Clustering is especially useful for gaining insights into human behaviour. For example, a company may use it to sort its customers into distinct groups, based on their purchase histories, so that it can target them more effectively with promotions.

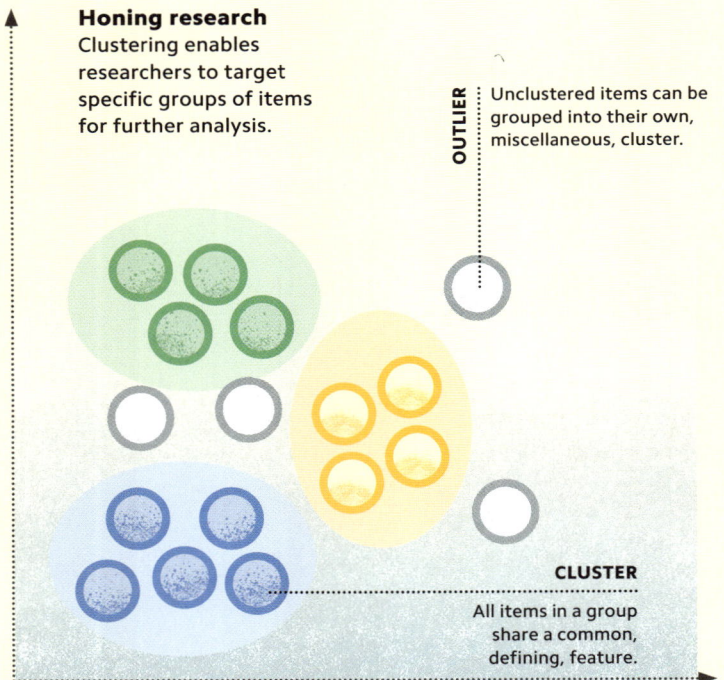

**Honing research**
Clustering enables researchers to target specific groups of items for further analysis.

**OUTLIER**
Unclustered items can be grouped into their own, miscellaneous, cluster.

**CLUSTER**
All items in a group share a common, defining, feature.

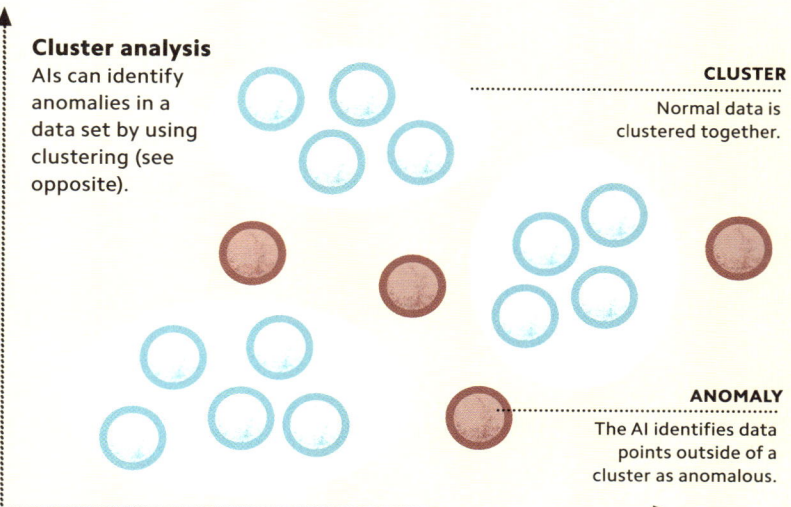

**Cluster analysis**
AIs can identify anomalies in a data set by using clustering (see opposite).

**CLUSTER**
Normal data is clustered together.

**ANOMALY**
The AI identifies data points outside of a cluster as anomalous.

# THE ODD ONE OUT

Anomaly detection is the process of identifying unusual (or "anomalous") data in a data set. That is to say that the AI looks for items that do not fit a particular pattern or model built from its training data. Many anomalies are caused by mistakes in the data, such as incorrectly inputted units, or an inconsistency in the type of measurement used. In such cases, it is important to find the anomaly so that it can be corrected or removed from the data set. However, anomalies can also draw attention to serious problems that lie outside the data set, such as a software malfunction or the AI being hacked by cybercriminals (see pp.96–97).

# THE MOST LIKELY OUTCOME?

Machine-learning models (see pp.58–59) can make predictions by analysing patterns in historical data. In AI, a prediction is the output from a model that forecasts the chances of a particular outcome. For example, if a customer buys a certain item online, an AI can use data about past purchases – both from the customer and from others – to predict what other items they may want. Prediction in AI does not always involve anticipating a future event. It can also be used to make "guesses" about events in the past and present, such as whether a transaction is fraudulent (see p.98), or if an X-ray indicates the presence of disease (see p.102).

**Customer purchase**
A customer purchases a product from an online vendor – for example, a toothbrush.

**Customer profile**
The AI builds a profile of a customer by analysing their online behaviour and history of purchases.

### Similar items

The AI identifies other items frequently bought alongside the product – both by the customer and others.

### Prediction

The AI predicts and then recommends linked items that the customer may want – for example, toothpaste and mouthwash.

### Similar profiles

The AI compares the customer's profile to a large number of other profiles to find similar matches.

### Prediction

The AI uses the purchase history of similar profiles to predict other items the customer might be interested in.

# MACHINE LEARNING WITH "LABELLED" DATA

Supervised learning is a type of machine learning in which an AI is trained using a "labelled" training data set (see p.61). Input and output data is labelled by a human so that the AI can learn the relationship between them. The inputs, outputs, and the rule that relates them are collectively known as a "function". During training, weights (see p.78) are adjusted to make the function fit the training data. The resulting function can be used to predict outputs based on new inputs. Supervised learning can be used for classification (see p.66) and regression (see p.67).

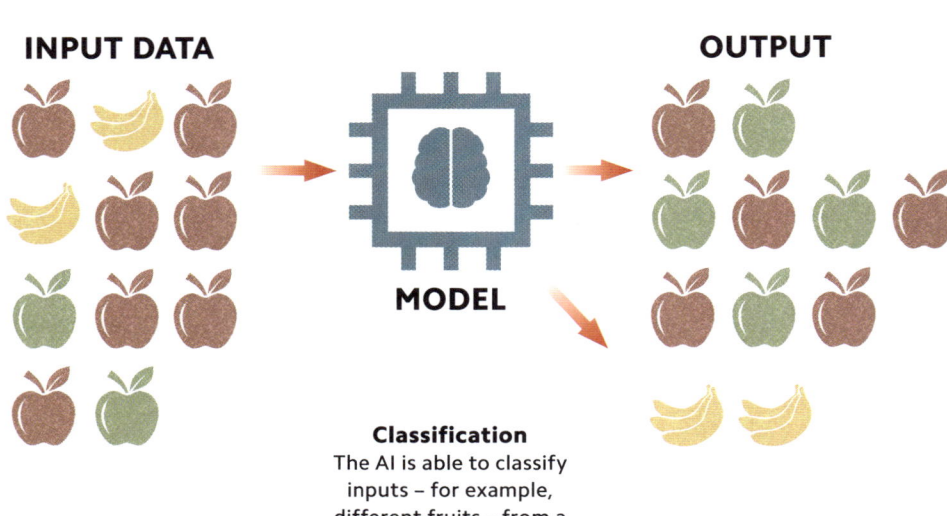

**INPUT DATA**

**MODEL**

**OUTPUT**

**Classification**
The AI is able to classify inputs – for example, different fruits – from a large, unlabelled data set.

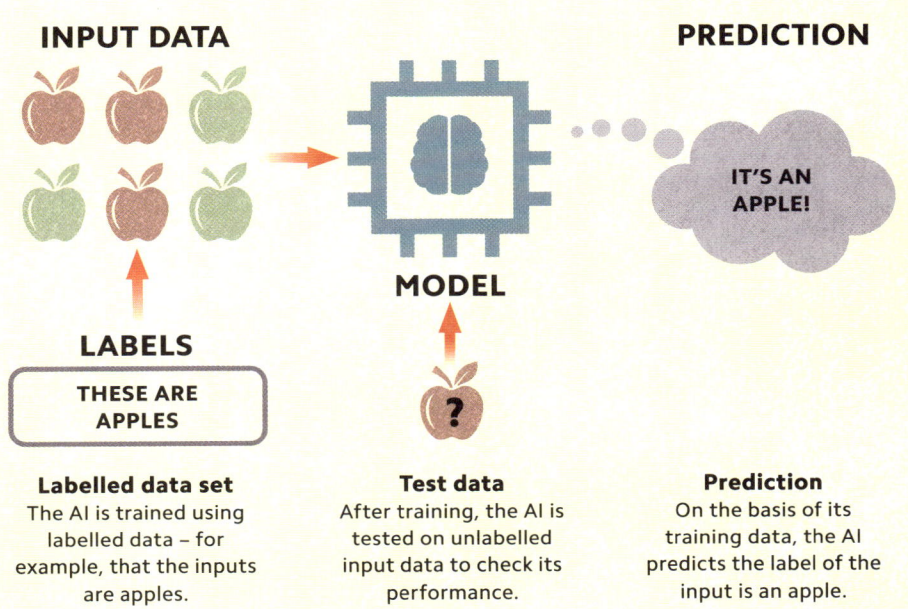

## INPUT DATA

## PREDICTION

IT'S AN
APPLE!

**MODEL**

## LABELS

THESE ARE
APPLES

**Labelled data set**
The AI is trained using
labelled data – for
example, that the inputs
are apples.

**Test data**
After training, the AI is
tested on unlabelled
input data to check its
performance.

**Prediction**
On the basis of its
training data, the AI
predicts the label of the
input is an apple.

# MACHINE LEARNING WITH "RAW" DATA

Unsupervised learning is used to discover hidden structures
in raw, unlabelled data sets. Although AIs do not understand
the relevance of these structures, they may still have real-world
meaning. This approach is useful in the early stages of data mining
(see p.60), to find patterns in large, unlabelled data sets, which can
then be subject to human interpretation. An in-between method,
semi-supervised learning, uses partly labelled data sets, which
gives better results than entirely unsupervised learning.

# LEARNING FROM FEEDBACK

Reinforcement learning is an approach to machine learning (see pp.58–59) in which an AI is taught to perform a task through trial and error. To achieve this, the AI is programmed to recognize "rewards" and "punishments", meaning positive or negative feedback, depending on whether it succeeds or fails. The AI learns that succeeding is good and failing is bad, and repeatedly attempts the task until it is rewarded. For example, an autonomous vehicle trained in this way (see p.122) will be punished – receive negative feedback – until it learns not to go through a red traffic light.

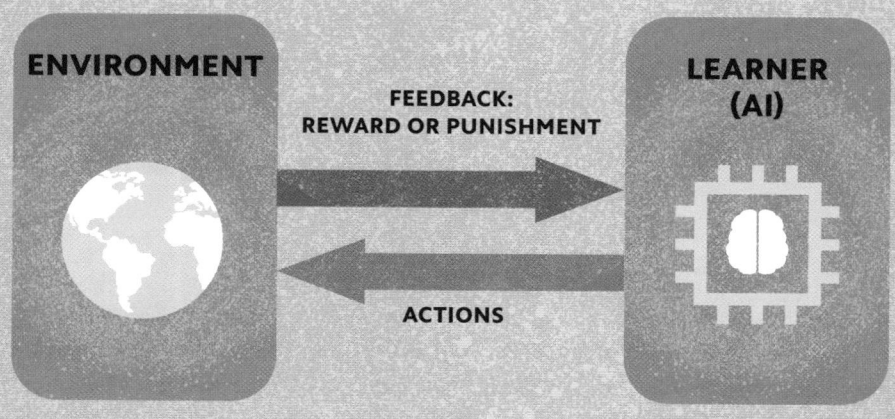

**ENVIRONMENT**

**FEEDBACK:
REWARD OR PUNISHMENT**

**LEARNER
(AI)**

**ACTIONS**

**Trial and error**
The AI learns to succeed in a
task through the consequences of its
actions. It will seek rewards and avoid
punishment until the task is completed.

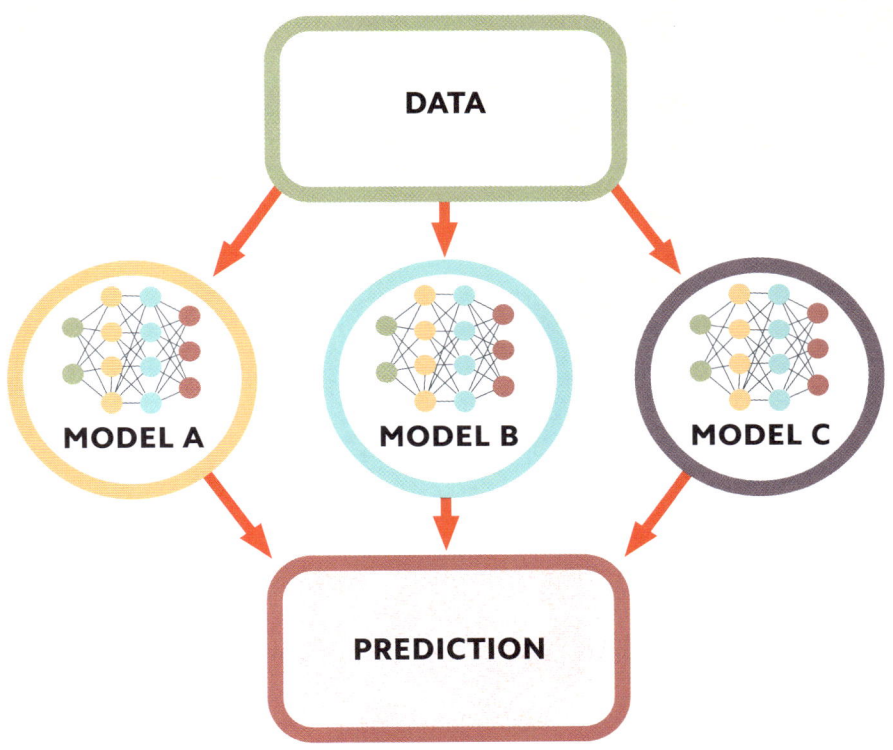

# WORKING TOGETHER

Ensemble learning is based on the idea that combining the outputs of multiple machine-learning algorithms produces a better result than a single model can. Using two or more models that have been built and trained in different ways, for example with different data sets, can "cancel out" their individual weaknesses and generate more accurate predictions. Ensemble learning can be used to "teach" a particular model to improve its predictive performance, but also to assess a model's reliability and prevent a poor one being selected.

**PROCESSING LAYERS**

Known as the "hidden layers", these are where data is processed within an ANN.

**OUTPUTS**

The output layer is the final layer, where useful results are obtained.

**INPUTS**

Information, such as data, enters an ANN via the input layer.

# THE AI BRAIN

Artificial neural networks (ANNs) are machine-learning models based on algorithms (see p.14). Their structure is similar to that of the brain, consisting of interconnected nodes – artificial neurons – that are organized into multiple "layers". The nodes within each layer receive, process, and send data to the next layer in the network, until an output, or result, is produced. Each node works like an individual microprocessor that can be reprogrammed to handle the data in a desired way. Using training data (see p.61), programmers can teach an ANN to "learn" how to give the expected results, or outcomes.

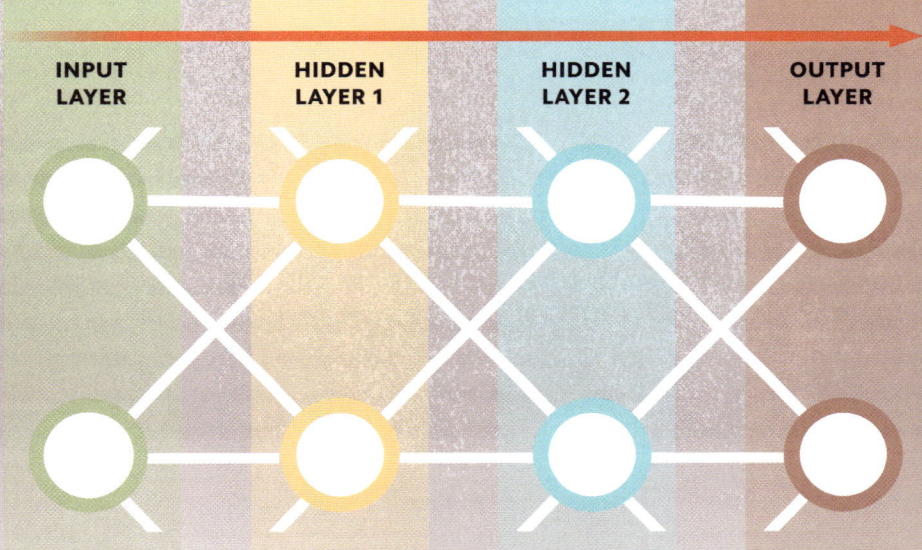

**INPUT LAYER**  **HIDDEN LAYER 1**  **HIDDEN LAYER 2**  **OUTPUT LAYER**

**Input layer**
The input layer brings the initial data into the network.

**Multiple hidden layers**
Data is processed within the "hidden" layers, passing through the network, from one layer to the next.

**Output layer**
The processed data leaves the network via the output layer.

# NETWORK STRUCTURE

Artificial neural networks (ANNs) are structured in "layers"– collections of processing nodes that operate together. Data flows from the nodes in one layer to those in the next. The first layer always contains the "input," or incoming data. Next is at least one "hidden" layer, in which the processing takes place. These layers are hidden in the sense that their data is not visible to a user in the way that the network's inputs and outputs are. Finally, the resulting data arrives at the "output" layer. All ANNs share this basic structure, but some are more complex: recurrent neural networks (see p.85) generate connections between nodes in the next or in previous layers, while deep neural networks (see p.86) can have hundreds of hidden layers.

# ASSIGNING IMPORTANCE

AI algorithms include variables – mathematical values that can change – that determine how data is processed within an artificial neural network (ANN, see pp.76). When designing and training an ANN, programmers can give these variables greater or lesser influence within the algorithm. This influence is known as "weight". The more weight an input has, the greater its influence over the output. The "bias" (see opposite) determines the threshold at which variables become significant. Adjusting the weights and bias allows the ANN to be fine-tuned to give more accurate results.

## SHOULD I SNACK ON AN APPLE?

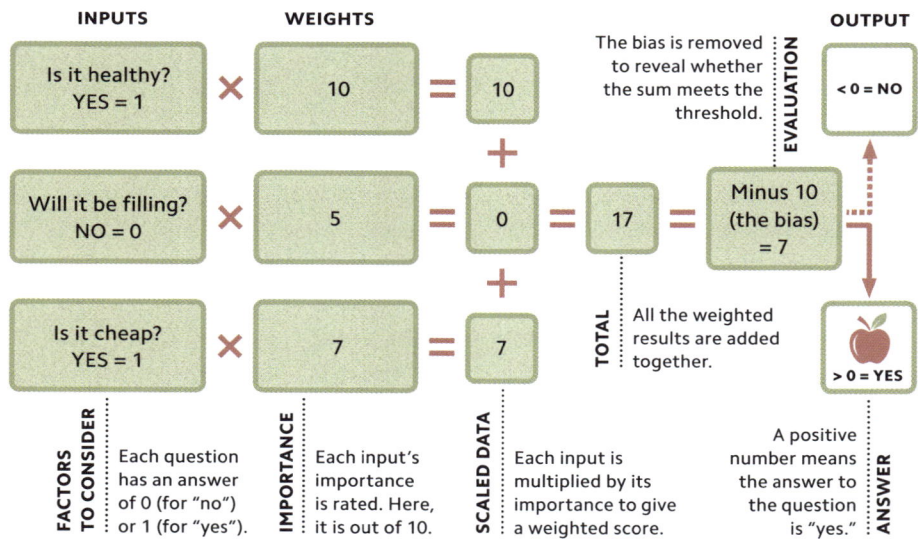

| | |
|---|---|
| **INPUTS** | **WEIGHTS** |

**INPUTS** **WEIGHTS** **OUTPUT**

Is it healthy? YES = 1  ×  10  =  10

The bias is removed to reveal whether the sum meets the threshold. **EVALUATION**

< 0 = NO

+

Will it be filling? NO = 0  ×  5  =  0  =  17  =  Minus 10 (the bias) = 7

+

Is it cheap? YES = 1  ×  7  =  7

> 0 = YES

**FACTORS TO CONSIDER**: Each question has an answer of 0 (for "no") or 1 (for "yes").

**IMPORTANCE**: Each input's importance is rated. Here, it is out of 10.

**SCALED DATA**: Each input is multiplied by its importance to give a weighted score.

**TOTAL**: All the weighted results are added together.

**ANSWER**: A positive number means the answer to the question is "yes."

# GOALS AND THRESHOLDS

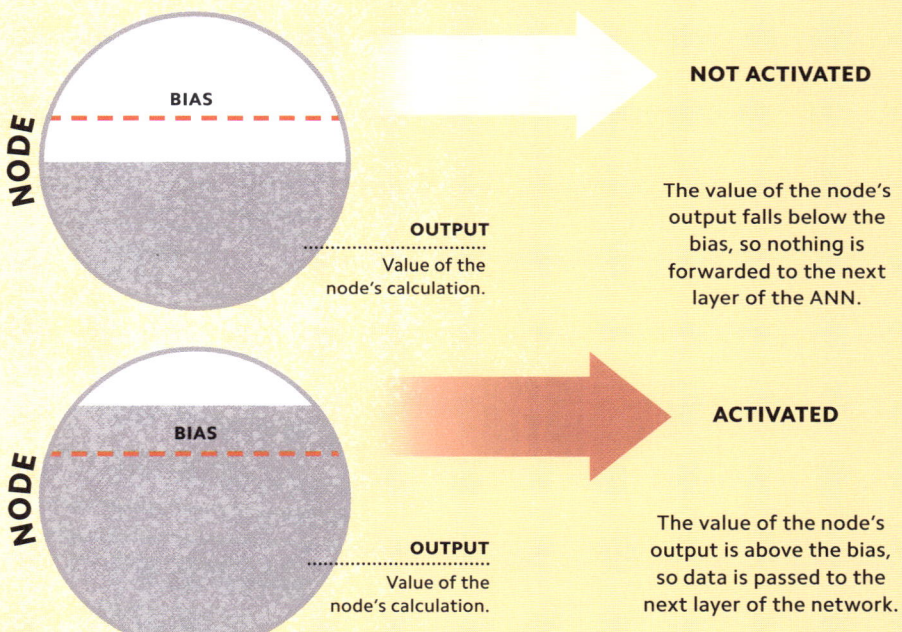

**NODE**

BIAS

OUTPUT
Value of the
node's calculation.

**NOT ACTIVATED**

The value of the node's
output falls below the
bias, so nothing is
forwarded to the next
layer of the ANN.

**NODE**

BIAS

OUTPUT
Value of the
node's calculation.

**ACTIVATED**

The value of the node's
output is above the bias,
so data is passed to the
next layer of the network.

An artificial neural network (ANN, see p.76) is made up of
layers of "nodes", which receive and process data. Before a
node can pass information on to the next layer of nodes,
its output data must reach a certain value. This value –
essentially a numerical score set by the ANN designer – is
known as the "bias". The node can only "activate" and pass
on its output data once the bias has been met. If the node
is not activated, that path of data transmission stops.
Different biases can also be set to direct data to specific
nodes on the next layer of the ANN.

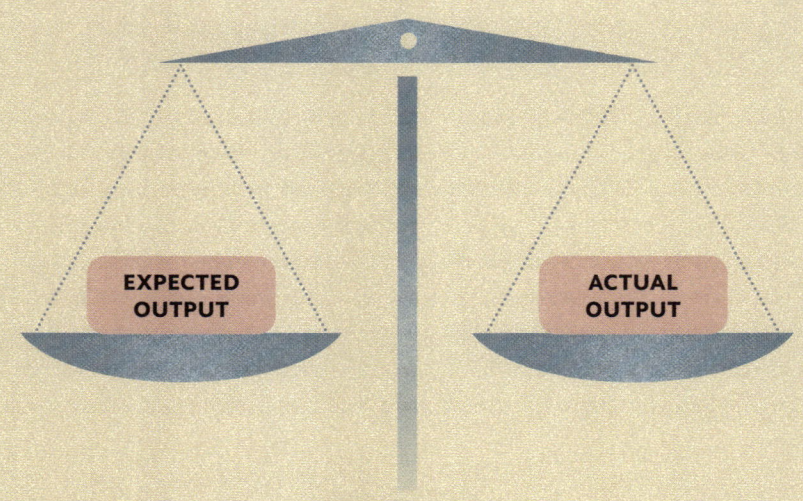

**Cost function**
The difference between the
expected and actual outputs gives
the model's performance. The goal
is for them to be equal.

# MEASURING SUCCESS

The performance of a machine-learning model, such
as an artificial neural network (see pp.76), can be evaluated
by its "cost function". This is a measure of the change that
occurs during training between the actual outputs from the
model and the outputs expected by the programmer. This
difference, called the "cost", is expressed as a number. The
higher the number, the greater the gap between the real and
the anticipated outputs, and the poorer the model. As the
model learns, the cost reduces and performance improves.
The training is complete when the cost is zero, or as close
to zero as possible.

# IMPROVING PERFORMANCE

A machine-learning model improves its performance by fine-tuning its settings. Instead of having to process huge amounts of data, the model can start at a random data point and then "nudge" its way towards a better solution. The algorithm that trains it to do this is known as the "gradient descent". Each time the model adjusts its settings, the gradient descent rates its success using the "cost function" (see opposite). Plotting the gradient of the cost function on a graph reveals a curve. The model reduces its cost function by following the steepest downward slope. When the slope levels off, the model is as good as it can be and it stops learning.

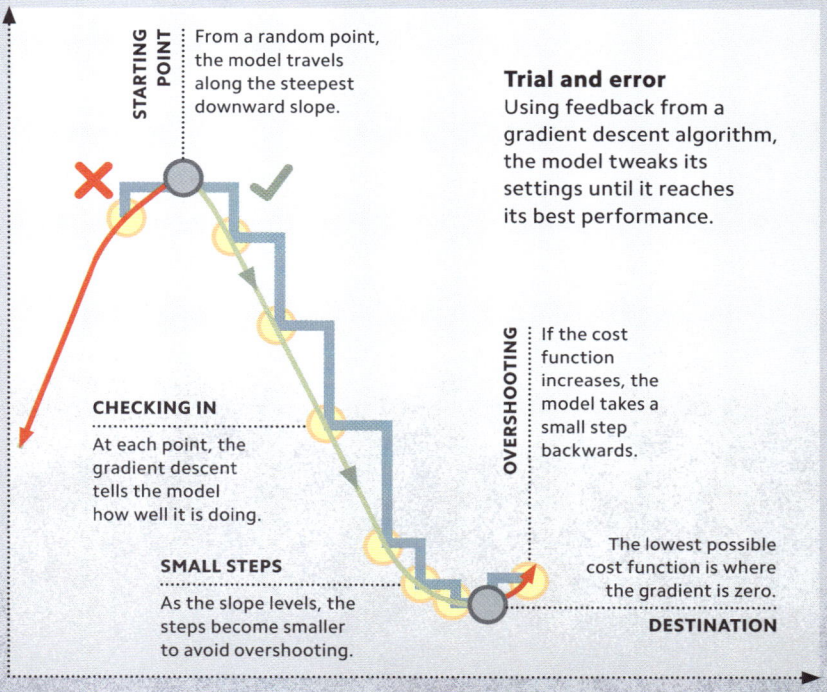

**STARTING POINT**
From a random point, the model travels along the steepest downward slope.

**Trial and error**
Using feedback from a gradient descent algorithm, the model tweaks its settings until it reaches its best performance.

**CHECKING IN**
At each point, the gradient descent tells the model how well it is doing.

**OVERSHOOTING**
If the cost function increases, the model takes a small step backwards.

**SMALL STEPS**
As the slope levels, the steps become smaller to avoid overshooting.

The lowest possible cost function is where the gradient is zero.

**DESTINATION**

# REFINING THE MODEL

**1. Test the model**
Run the ANN using training data (see p.61). The output is called the "test data".

**2. Calculate the cost function**
Compare the test data to the training data. The difference is the cost function (see p.80).

The "delta rule" – also known as the delta learning rule – enables a single-layer artificial neural network (ANN, see pp.76) to boost its performance by refining its settings. It makes use of "gradient descent" (see p.81) to identify the best choices for improving the model. As the model's output gets nearer to the expected output, smaller adjustments are carried out, until the outputs are as close as possible to each other. Backpropagation (see p.84) is a generalized form of the delta rule that applies to ANNs with any number of layers.

**4. Update the model**
Amend the settings in the ANN based on the feedback from the gradient descent.

**3. Use a gradient descent algorithm**
This determines which direction the model's settings should move for a lower cost function.

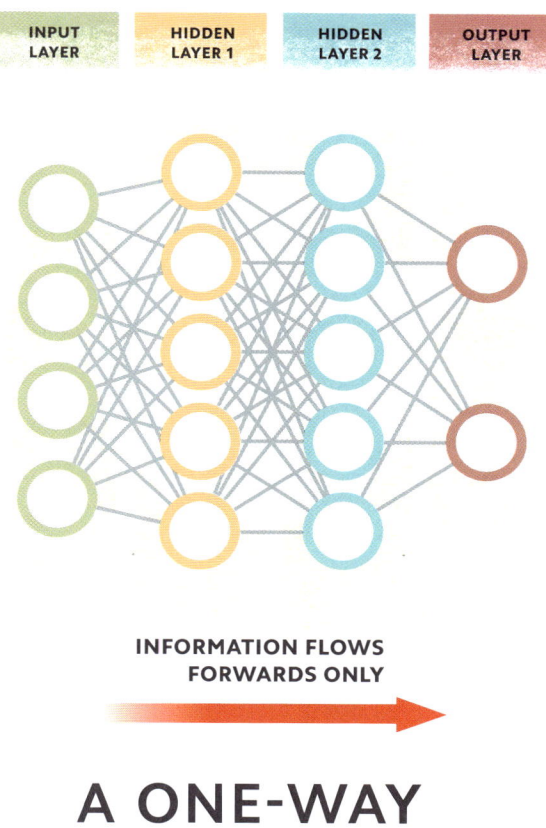

**INFORMATION FLOWS FORWARDS ONLY**

# A ONE-WAY NETWORK

A "feedforward neural network" (FNN) is a simple artificial neural network (ANN, see p.76) in which information flows forwards only – from the input layer, through the hidden layers, to the output layer. The connections between the nodes in an FNN do not form "feedback loops" – in other words, outputs are not fed backwards as inputs, as they are in a recurrent neural network (RNN, see p.85). The most basic form of feedforward neural network is a single artificial neuron (see p.21), which can undergo machine learning using gradient descent (see p.81).

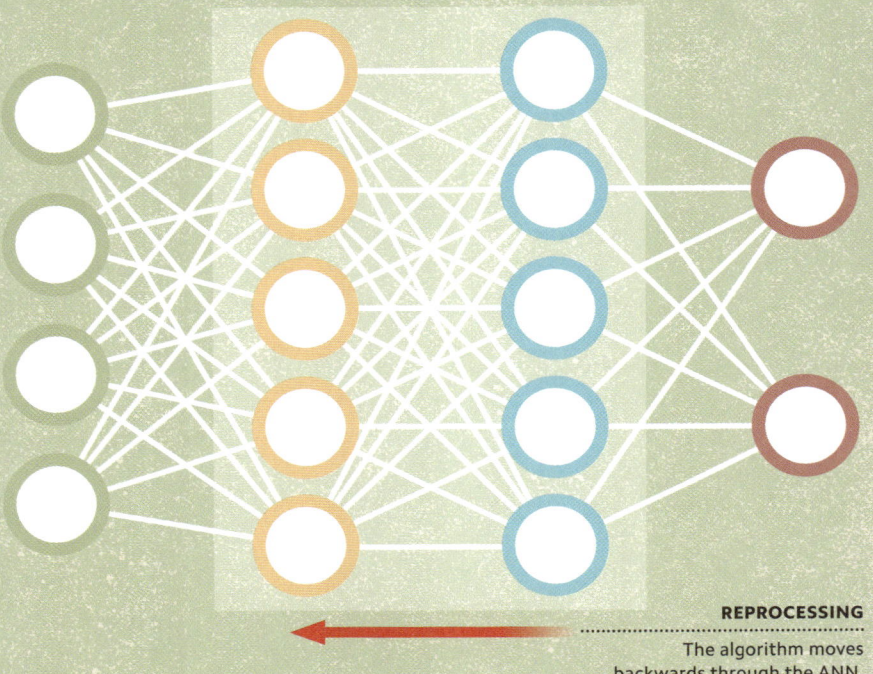

**REPROCESSING**

The algorithm moves backwards through the ANN, fine-turning it layer by layer.

# FINE-TUNING DATA

Backpropagation is a type of algorithm that is used to train artificial neural networks (ANNs, see p.76), specifically feedforward neural networks (see p.83). It is known as backpropagation because it begins at the final (output) layer and moves in reverse towards the first (input) layer. During this process, nodes are reprogrammed by adjusting their weights (see p.78) and biases (see p.79), using gradient descent (see p.81) to find out whether increasing or decreasing them will produce better results. This has the effect of fine-tuning the ANN to produce more accurate outcomes overall.

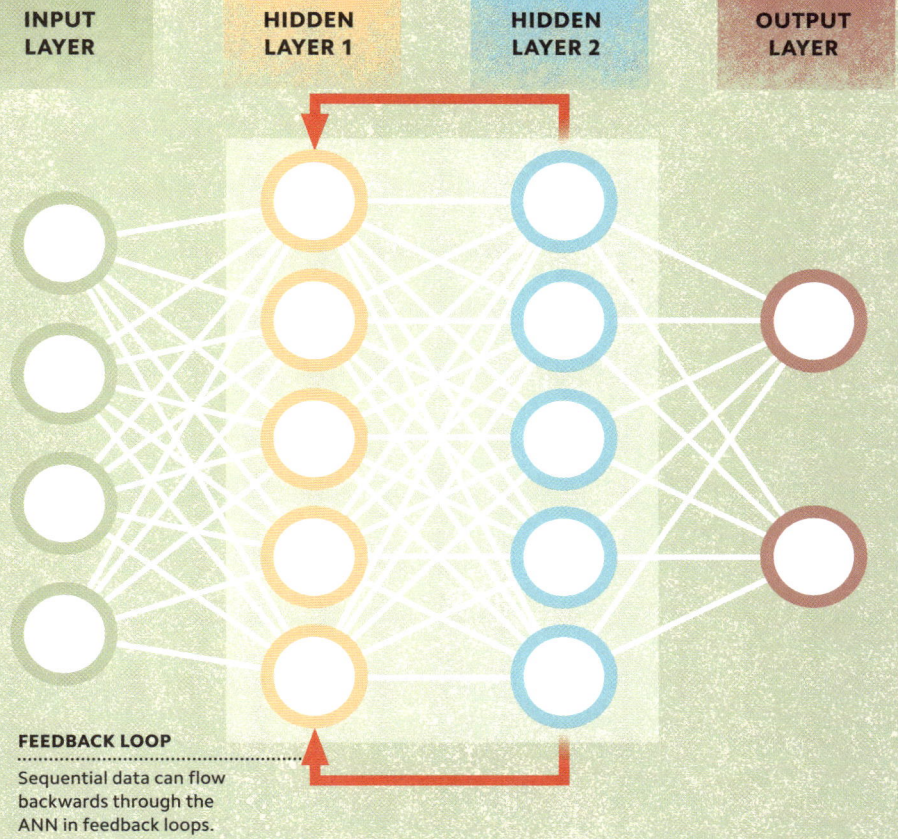

INPUT
LAYER

HIDDEN
LAYER 1

HIDDEN
LAYER 2

OUTPUT
LAYER

**FEEDBACK LOOP**
Sequential data can flow
backwards through the
ANN in feedback loops.

# STRUCTURED DATA

A recurrent neural network (RNN) is a type of ANN in which data
can move backwards in a "feedback loop". RNNs are used to process
sequential data – data that has to be in a specific order – such as
language. While traditional ANNs process individual data points to give
an outcome, RNNs maintain the essential structure and relationships
within sequential data, so that it remains intact. In doing so, RNNs
can be used to predict the next output of a sequence. They are used
widely in natural language processing tasks (see pp.112–13), including
training virtual assistants to carry out spoken conversations.

OUTPUT LAYER

MULTIPLE LAYERS

Deep neural networks have many layers of nodes to complete complex tasks.

HIDDEN LAYERS

INPUT LAYER

# BUILDING A BRAIN

Deep learning is a powerful form of machine learning based on artificial neural networks (ANNs, see p.76). It uses ANNs with many hidden layers, known as deep neural networks (DNNs), to identify increasingly more meaningful features from input data. As with ANNs, data passes from the input layer into the hidden layers, where the nodes receive, process, and pass it on to another node in the next layer. With so many layers, DNNs process data very accurately and quickly, and are also capable of making accurate predictions. They are used in many complex AI processes, such as natural language processing (see pp.112–13).

# AI VS AI

A generative adversarial network (GAN) is a machine-learning model that uses two competing artificial neural networks (ANNs, see p.76). One ANN, the "generator", uses unlabelled training data (see p.61), supplied by the programmer, to create new fake data that it supplies to the second ANN – the "discriminator". The discriminator aims to identify the fake data. If it succeeds, the generator tries again, creating fake data that is harder to distinguish from real data. If the discriminator fails, it tries again to identify fake data more effectively. This process continues until the generator can make convincing fake data. In other words, it can use existing data to create accurate new data, such as predictions.

**Machine learning**
The generator's fake data is not convincing enough. It tries again, gradually improving the quality of its fakes.

**Spotted**
Fake data is identified.

**Not spotted**
Fake data is not identified.

**Machine learning**
The discriminator has not identified the fake data. It tries again until it succeeds, learning in the process.

**DISCRIMINATOR**
This ANN tries to identify the fake data within the real data.

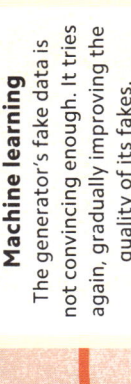

**GENERATOR**
This ANN produces fake data to trick the discriminator.

# PROCESSING VISUAL DATA

A convolutional neural network (CNN) is a type of deep neural network (see p.86) that is similar to the structure of the visual cortex: the part of the brain that takes and analyses information from the eye. CNNs are effective tools for computer vision (see p.110), since they can be taught to recognize features in input images, such as the pointed ears of cats. There are three types of layers (see p.77) in a CNN. The first type performs a function called a "convolution", which allows features

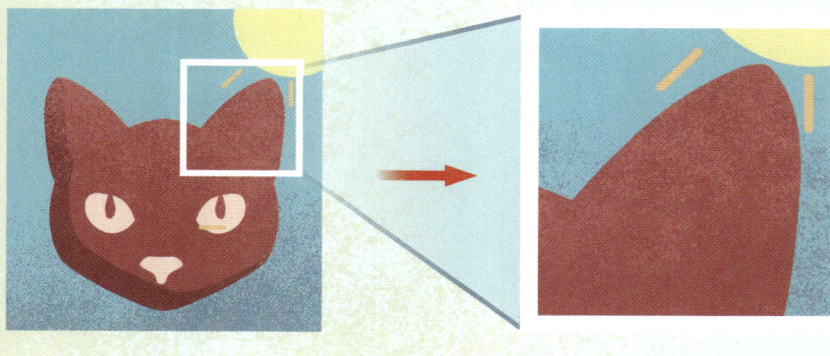

**Input**
The input in CNNs is typically an image – such as a photograph of a cat.

**Convolution**
A filter is applied to the image to produce feature maps. This enables features to be detected in patterns of pixels.

"I get very excited when we discover a way of making neural networks better – and when that's closely related to how the brain works."
Geoffrey Hinton

in an image to be detected. These layers first extract low-level features (lines and edges), before extracting higher-level features (shapes). They work by passing a filter over the image that creates a "map" of the location of each feature on the image. Between each convolution, there is a "pooling" layer, which reduces the complexity of the feature maps. The data from these layers is flattened, and then passes through a "classification" layer (see p.66), which identifies and labels the image.

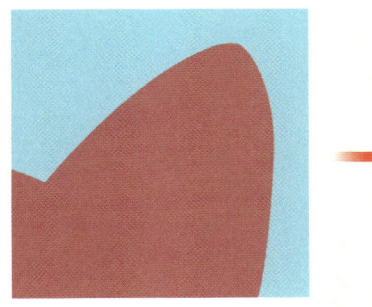

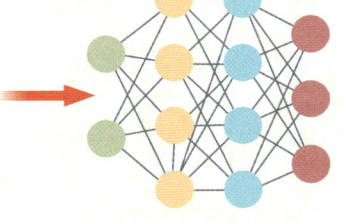

**Pooling**
"Mess" is cut out to reduce the amount of computing power required, and the features are abstracted.

**Classification**
Through a process of classification, the AI associates the data from the previous layers with an image.

**Output**
The AI identifies the photograph as being one of a cat.

# USING

# ARTIFIC

# INTELLI

# IAL
# GENCE

**Like computing**, AI has become a general-purpose technology, and has uses in a wide range of fields, from fine art to hi-tech weapons design. It is most effective when it is used as a tool to assist – rather than replace – human experts, and when very large quantities of data are involved, as with the Internet of Things (IOT). At its best, it can perform tasks with superhuman speed and accuracy. Sometimes, just one AI technique is required to perform a task, while other applications use a combination of techniques. For instance, autonomous vehicles incorporate many AI techniques from different fields, including computer vision, coupled with sonar, radar, and GPS technologies.

## USES OF AI
### (NOT EXHAUSTIVE)

**SEARCHING**
- RANKING (SEE P.94)
- RECOMMENDING (SEE P.95)

**DETECTING THREATS (SEE P.96)**

**FINANCE**

**RESEARCH**
- UNRAVELLING PROTEINS (SEE P.100)
- SEARCHING FOR PLANETS (SEE P.101)

**MEDICAL**
- DIGITAL DOCTORS (SEE P.102)
- MONITORING HEALTH (SEE P.103)

**INTERNET OF THINGS (SEE P.104)**
- MONITORING SYSTEMS (SEE P.106)
- "SMART" FARMING (SEE P.107)

**SMART DEVICES (SEE P.105)**

**SENSORY AI (SEE P.108)**

**UNDERSTANDING WORDS (SEE PP.112–113)**
- AI INTERPRETERS (SEE P.114)
- TALKING WITH AI (SEE P.115)
- AI HELPERS (SEE P.116)

**AI ARTISTS (SEE P.117)**

**INTELLIGENT ROBOTS (SEE P.118)**

ONLINE ATTACKS
(SEE P.97)

DETECTING FRAUD
(SEE P.98)

AI IN FINANCE
(SEE P.99)

PROCESSING SOUND
(SEE P.109)

MIMICKING SIGHT
(SEE P.110)

FACIAL RECOGNITION
(SEE P.111)

AI COMPANIONS
(SEE P.119)

MOVEMENT AND
MOBILITY (SEE P.120)

MANUAL DEXTERITY
(SEE P.121)

DRIVERLESS CARS
(SEE P.122)

AI AND WARFARE
(SEE P.123)

"A world run by automatons doesn't seem completely unrealistic anymore."
Gemma Whelan

# USES OF AI

Whether via mobile phones or virtual assistants, we interact with AI applications everywhere – often without realizing it. Using these applications has changed the way we work, shop, and communicate, and has revolutionized many industries, including finance, healthcare, and agriculture. Other AI technologies, such as generative AI and autonomous weaponry, are still in their infancy, but these too will soon be in mainstream use.

# RANKING

When an internet search engine is used, AI-generated rankings determine which sites appear most highly in the results. Some ranking algorithms locate and rank websites that contain the same terms, or "keywords", as those entered into the search engine by a user. Those with the closest matches rank the highest. Other algorithms rank websites more highly if they are accessed from many other sites, or if they are especially popular.

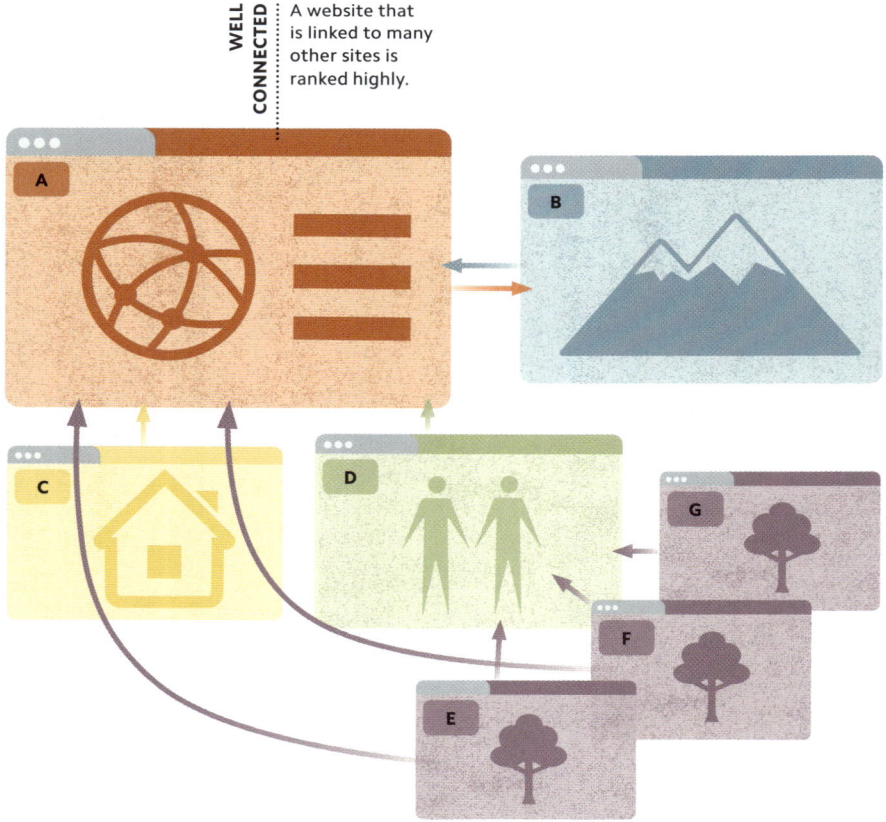

WELL CONNECTED
A website that is linked to many other sites is ranked highly.

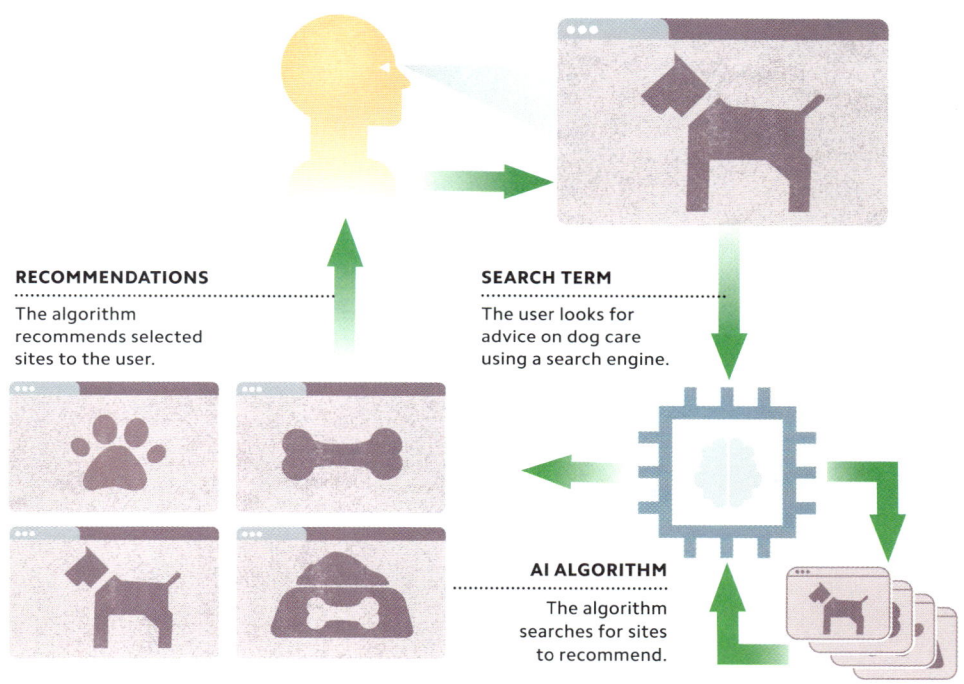

**RECOMMENDATIONS**

The algorithm recommends selected sites to the user.

**SEARCH TERM**

The user looks for advice on dog care using a search engine.

**AI ALGORITHM**

The algorithm searches for sites to recommend.

# RECOMMENDING

Based on an internet user's browsing history, and that of others, AI recommendation algorithms can suggest websites, as well as products, that may interest the user. This can involve suggesting similar content to what the user has viewed previously or offering sites that similar web users have visited. To do this, algorithms make predictions (see pp.70–71). For example, if an internet user searches for advice on dog care, the AI algorithm will predict that they have or want a dog. It then searches the internet to find popular sites, and products, associated with dogs.

# DETECTING THREATS

Traditional threat-detection software used by cybersecurity experts searches for known malware (see opposite) "signatures", blocks the malware files it detects, and raises alerts. Incorporating an AI into the system enables cyber defences to identify and categorize new and mutated threats ("zero-day" malware) that would otherwise be undetectable since they do not match any known signatures. This is a vital development in cybersecurity, given the speed with which new threats arise. AIs are also used to predict how and where a system might be breached, and to help respond to breaches.

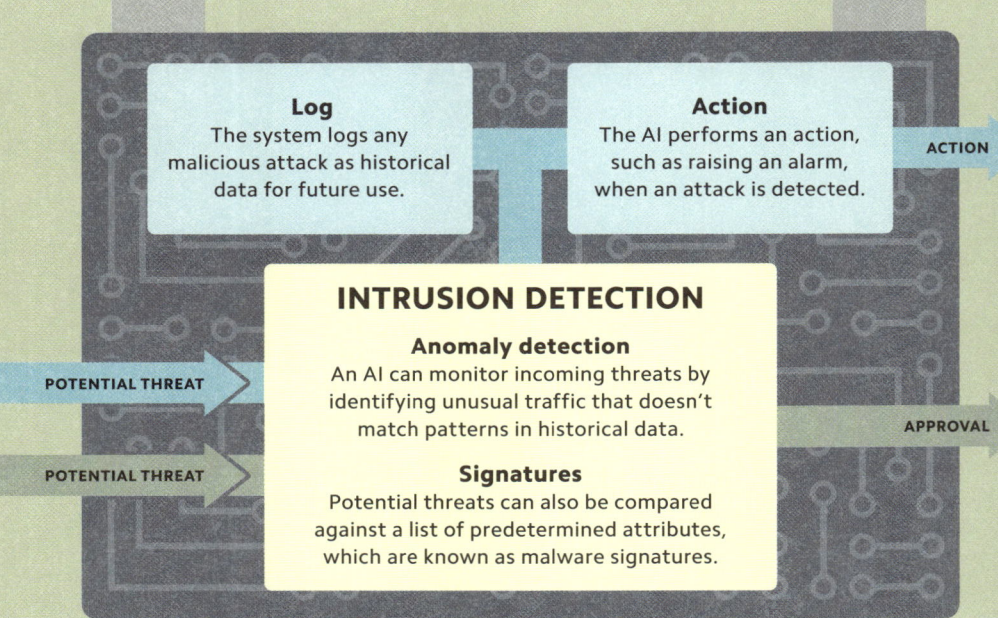

**Log**
The system logs any malicious attack as historical data for future use.

**Action**
The AI performs an action, such as raising an alarm, when an attack is detected.

ACTION

## INTRUSION DETECTION

### Anomaly detection
An AI can monitor incoming threats by identifying unusual traffic that doesn't match patterns in historical data.

### Signatures
Potential threats can also be compared against a list of predetermined attributes, which are known as malware signatures.

POTENTIAL THREAT

POTENTIAL THREAT

APPROVAL

## Hacking

Breaking into a system or device in order to access other people's digital information is known as "hacking".

## Ransomware

A type of malware, ransomware is designed to find and encrypt files on a device, making them inaccessible until a ransom is paid.

## Malware

Short for "malicious software", malware is any computer program that damages a device or gains access to sensitive information.

## Denial-of-service

A denial-of-service (DoS) attack floods a server with data, overwhelming it to the point at which it can no longer function.

## Disinformation

Using the internet, enemy agents can spread false news stories to influence public opinion, create instability, and stir up social unrest.

# ONLINE ATTACKS

The use of cyberattacks to target a nation state is known as "cyberwarfare". It is possible to inflict serious harm on a country remotely, disrupting key services and critical infrastructure such as power grids by disabling the information systems that control them. Cyberwarfare tactics include denial-of-service (DoS) attacks, malware such as viruses and ransomware, disinformation campaigns, and state-sponsored hacking. AI is used in cyberwarfare to enhance these attacks, making them faster and more sophisticated. For example, AI-driven malware is very hard to detect: it is able to use machine learning (see p.58–59) to find weaknesses in a device's security system, attack it while posing as an accidental error, and then cause harm to the device.

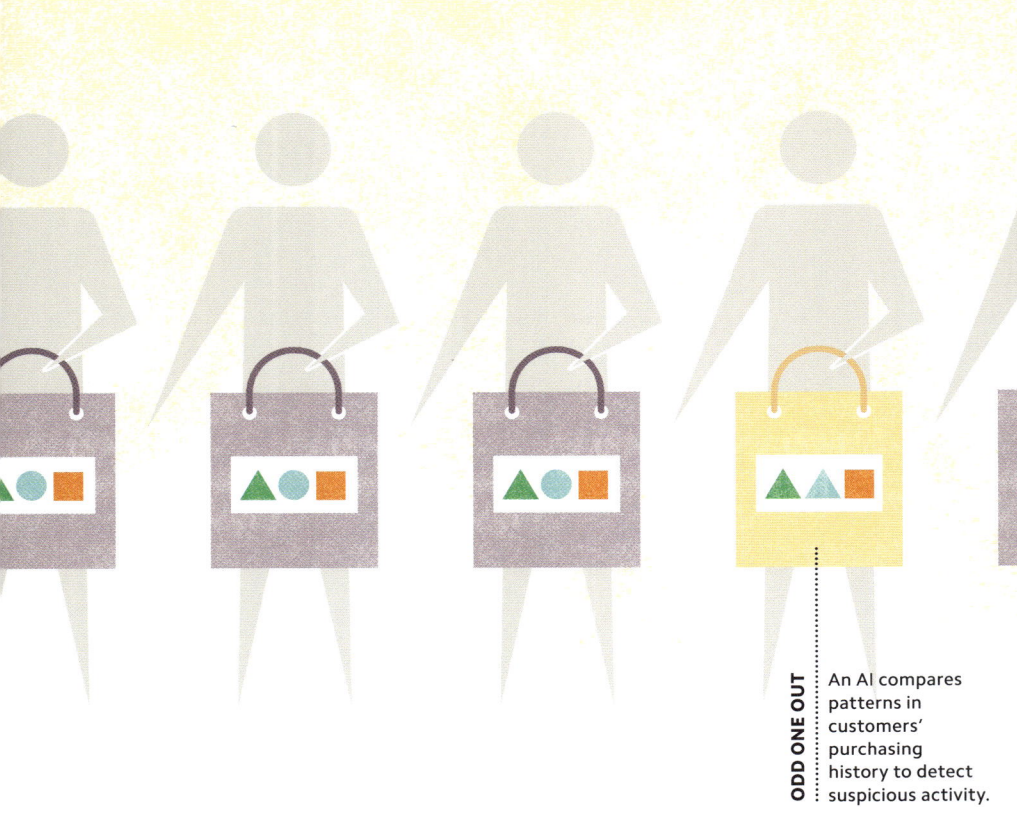

An AI compares patterns in customers' purchasing history to detect suspicious activity.

# DETECTING FRAUD

Financial institutions are adopting AI systems to detect – and prevent – fraud. These systems can process vast amounts of data about past transactions, learning the ordinary patterns of behaviour of a bank's customers. When transactions are made that do not fit this pattern (see p.69), an AI may flag them up as needing to be investigated or take other actions, such as freezing the customer's account. An AI may score each transaction on its likelihood of being fraudulent, then raise an alert when this score exceeds a certain threshold.

# AI IN FINANCE

High-frequency trading (HFT) is the use of specialized algorithms to make investment decisions and transactions at superhuman speed – performing millions of trades each day. Some financial institutions manage entire investment portfolios using HFT. By evaluating vast quantities of market data in real time, it can identify the best stocks and shares to buy and sell, the optimal time to place those deals, and perform transactions extremely quickly. To help inform its decisions, HFT may use natural language processing (NLP, see pp.112–113) to analyse news reports and social media.

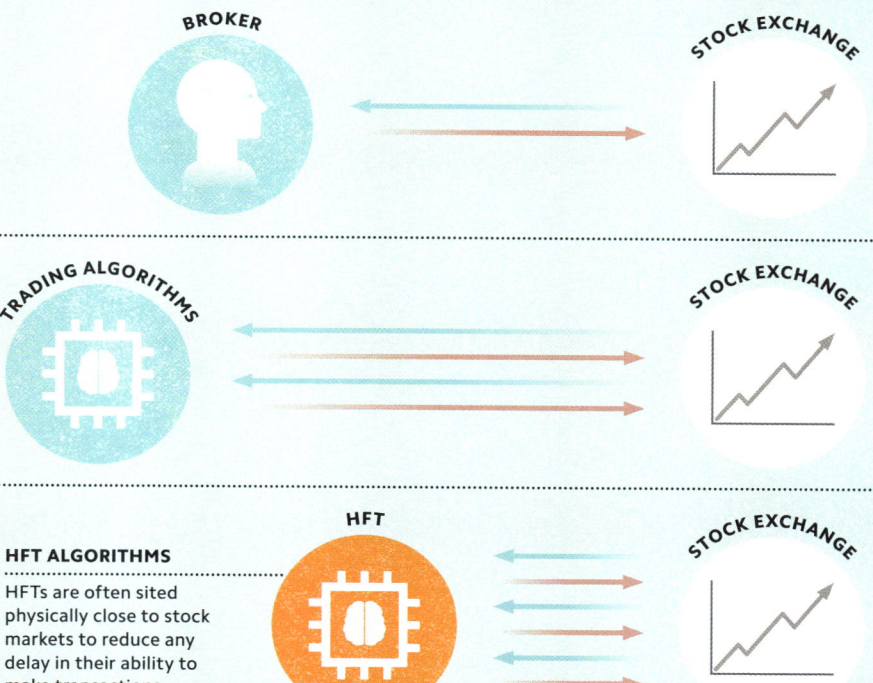

**BROKER**

**STOCK EXCHANGE**

**TRADING ALGORITHMS**

**STOCK EXCHANGE**

**HFT ALGORITHMS**

HFTs are often sited physically close to stock markets to reduce any delay in their ability to make transactions.

**HFT**

**STOCK EXCHANGE**

# UNRAVELLING PROTEINS

AIs not only speed up tedious work, they help to open up new fields of scientific research. For example, using deep learning (see p.86) and painstakingly collected experimental data, scientists have taught AIs to predict the 3D structure of "folded proteins" – the building blocks of life – with atomic precision. Previously, scientists could not tell how a protein's chemistry determined its folded structure. This "protein folding problem" was so complex that it remained unsolved for decades. Today, understanding how these proteins work has transformed medical research and accelerated the process of developing new drugs.

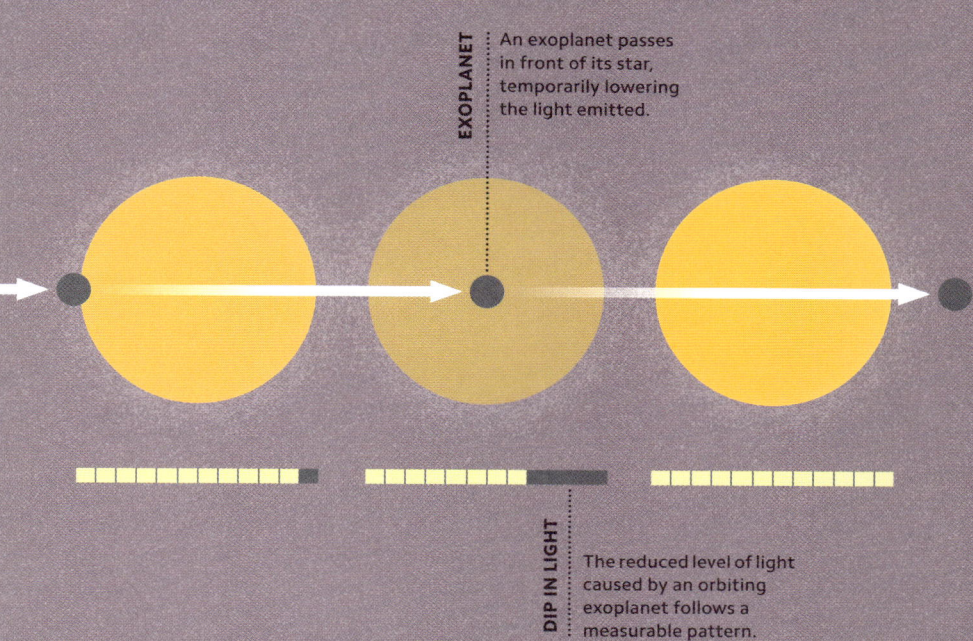

**EXOPLANET**
An exoplanet passes in front of its star, temporarily lowering the light emitted.

**DIP IN LIGHT**
The reduced level of light caused by an orbiting exoplanet follows a measurable pattern.

# SEARCHING FOR PLANETS

AI is a powerful tool in scientific research, enabling scientists to look for interesting phenomena in enormous quantities of data. For example, in astronomy, AIs are used to classify galaxies, look for gravitational waves, and identify "exoplanets" with high accuracy. An exoplanet is any planet outside our solar system. By measuring how much of a star's light is blocked over time, an artificial neural network (see p.76) can recognize whether or not this pattern is caused by an orbiting exoplanet. Hundreds of exoplanets have been discovered using AIs in this way.

# DIGITAL DOCTORS

AI is fast becoming a powerful tool for assisting doctors. Machine learning, and especially deep learning (see p.86), has proven effective at identifying disease in medical imagery, including finding signs of lung cancer on CT scans and detecting retinal problems caused by diabetes using photographs of patients' eyes. AI is also used to identify people at high risk of certain conditions, prioritize urgent cases, and help doctors to select treatments.

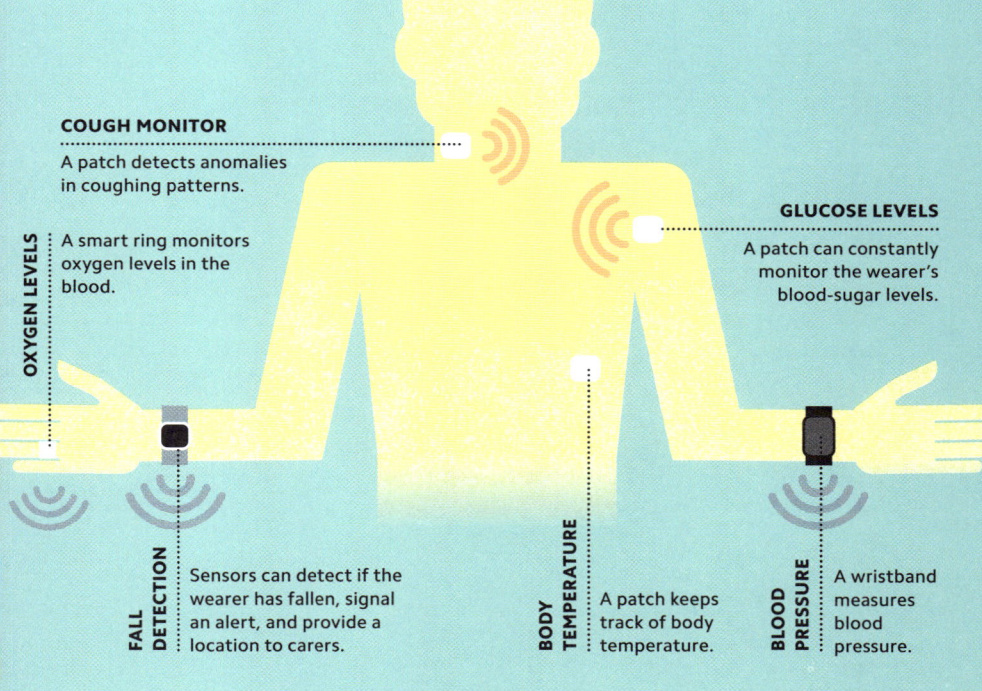

**COUGH MONITOR**
A patch detects anomalies in coughing patterns.

**GLUCOSE LEVELS**
A patch can constantly monitor the wearer's blood-sugar levels.

**OXYGEN LEVELS**
A smart ring monitors oxygen levels in the blood.

**FALL DETECTION**
Sensors can detect if the wearer has fallen, signal an alert, and provide a location to carers.

**BODY TEMPERATURE**
A patch keeps track of body temperature.

**BLOOD PRESSURE**
A wristband measures blood pressure.

# MONITORING HEALTH

AIs perform an important role in a new field of medicine known as "telehealth". By wearing sensors that monitor vital bodily functions, such as oxygen intake and blood pressure, a person can go about their day knowing that if a sensor detects a problem, it will send a signal to their digital assistant (an app on their phone or personal computer), which in turn – via the internet – will alert an AI at a healthcare centre. This AI will then compare the digital assistant's report with previous data about the person and alert a physician if necessary. Crucially, this technology can detect problems that a person may not even be aware of. More generally, AI technologies can also be used to monitor people's general fitness and wellbeing.

# INTERNET OF THINGS

The "Internet of Things" (IOT) is the network of interconnected devices that collect and exchange data via the internet – not only phones and computers, but also smart fridges, driverless cars, fitness monitors, security cameras, and tens of billions of other items. Due to the vast amounts of data that these devices collect – and the requirement that they respond appropriately to their users and environment – AIs have become integral to the IOT. A smart energy meter, for example, may use an AI to identify patterns in a user's energy consumption and suggest adjustments to reduce their bills.

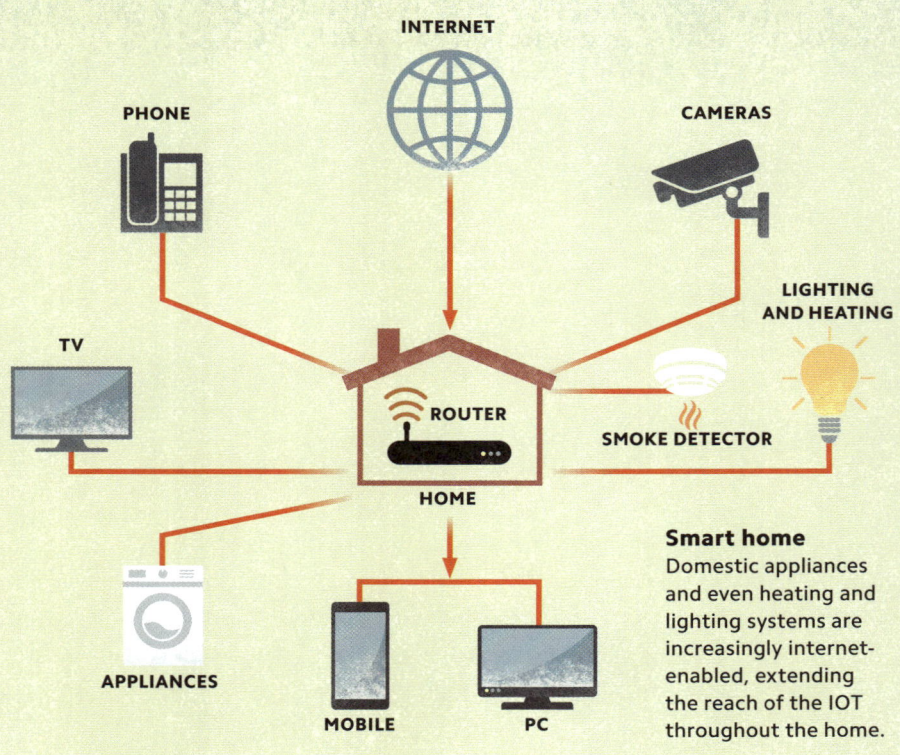

**INTERNET**

**PHONE**

**CAMERAS**

**LIGHTING AND HEATING**

**TV**

**ROUTER**

**SMOKE DETECTOR**

**HOME**

**APPLIANCES**

**MOBILE**

**PC**

**Smart home**
Domestic appliances and even heating and lighting systems are increasingly internet-enabled, extending the reach of the IOT throughout the home.

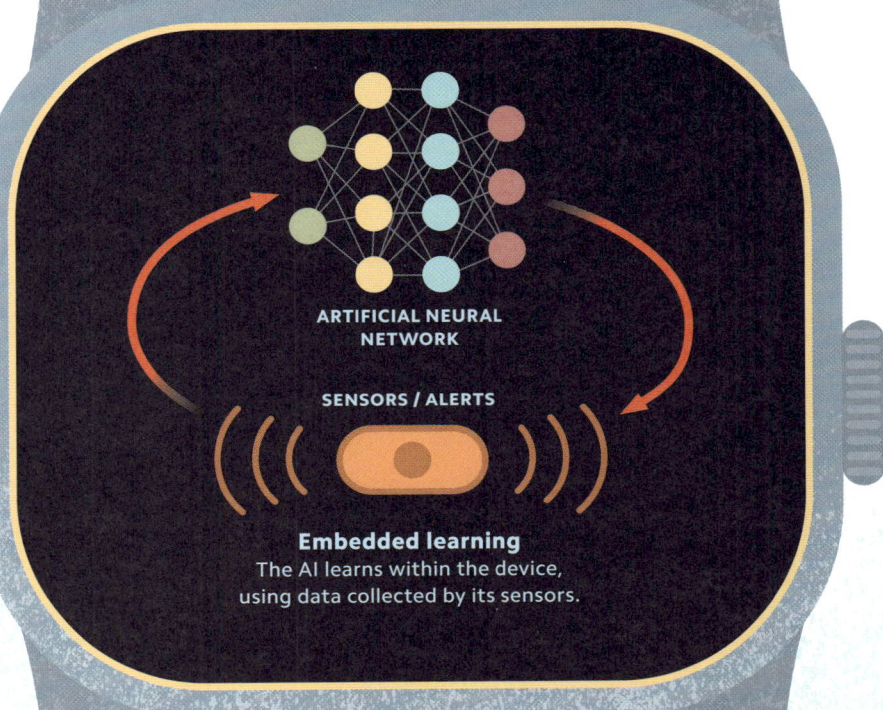

**ARTIFICIAL NEURAL NETWORK**

**SENSORS / ALERTS**

**Embedded learning**
The AI learns within the device,
using data collected by its sensors.

# SMART DEVICES

The "intelligence" in the Internet of Things (IOT,
see opposite) is mostly contained within clouds –
remote computing systems usually owned by
technology companies. Increasingly, however, AI
software capable of machine- and deep learning is
being embedded in devices, such as mobile phones
and smart watches. Using embedded AI removes
the need to send data to and from the cloud
continuously, reducing power usage, data
processing time, risk of data breaches, and reliance
on cloud providers. In real-time monitoring devices
(see p.106), embedded AI allows almost instant
detection and response.

# MONITORING SYSTEMS

The "Internet of Things" (see p.104) enables AIs to monitor all kinds of equipment automatically, up to major infrastructure systems, such as gas pipelines, transport networks, and electricity grids. Sensors distributed throughout these systems collect and transmit their data to AIs, which then scan the data for anomalies (see p.69) and alert human technicians to investigate them further if necessary. AIs are also used to predict where faults could occur in the future, enabling technicians to take action to prevent equipment failure. Such measures minimize the disruption caused by using complex equipment that needs regular maintenance.

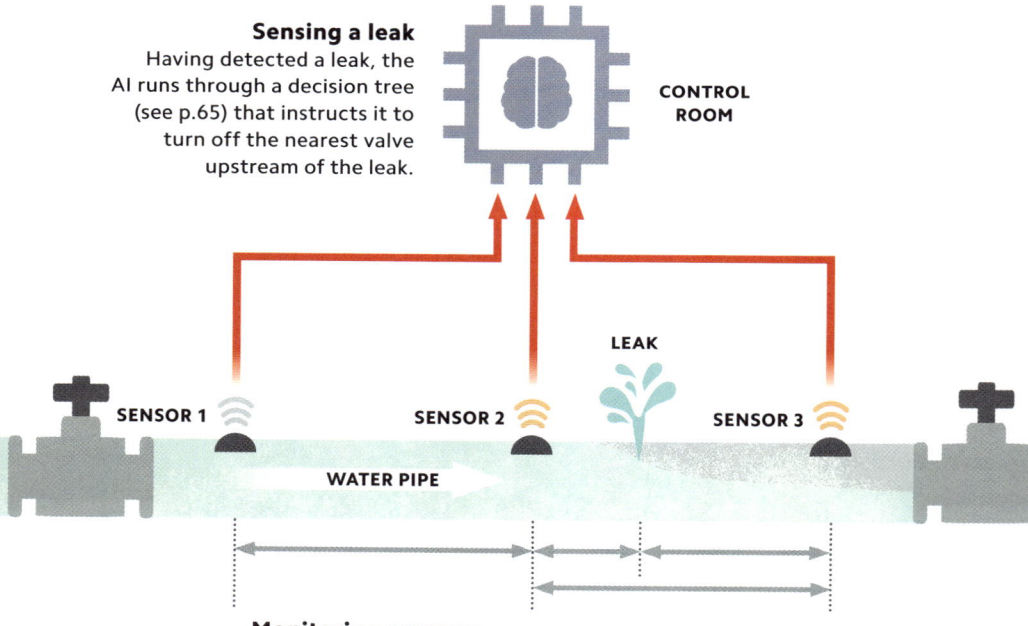

**Sensing a leak**
Having detected a leak, the AI runs through a decision tree (see p.65) that instructs it to turn off the nearest valve upstream of the leak.

CONTROL ROOM

LEAK

SENSOR 1

SENSOR 2

SENSOR 3

WATER PIPE

**Monitoring pressure**
Sensors monitor the pressure within a water pipe and wirelessly transmit their data to an AI. Here, the AI detects an anomaly; the pressure is lower than it should be between sensors 2 and 3.

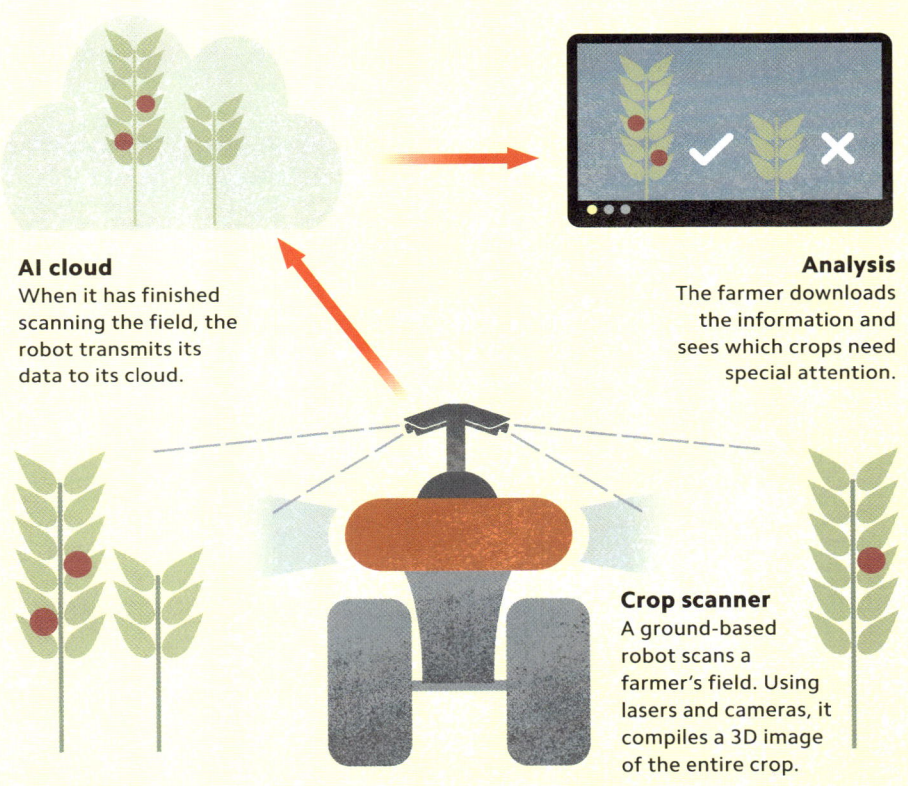

**AI cloud**
When it has finished scanning the field, the robot transmits its data to its cloud.

**Analysis**
The farmer downloads the information and sees which crops need special attention.

**Crop scanner**
A ground-based robot scans a farmer's field. Using lasers and cameras, it compiles a 3D image of the entire crop.

# "SMART" FARMING

AI is a key technology in "precision agriculture" – an approach to farming that optimizes the use of water and other resources in order to increase yields and minimize waste. Using devices such as drones in the air and robots on the ground, which collect data analysed by AIs, farmers can receive real-time information about their crops, enabling them to know which of them require water, pesticide, or fertilizer at any time. Such precise methods of farming may become indispensable in the coming decades, when the global population is set to increase by two billion people.

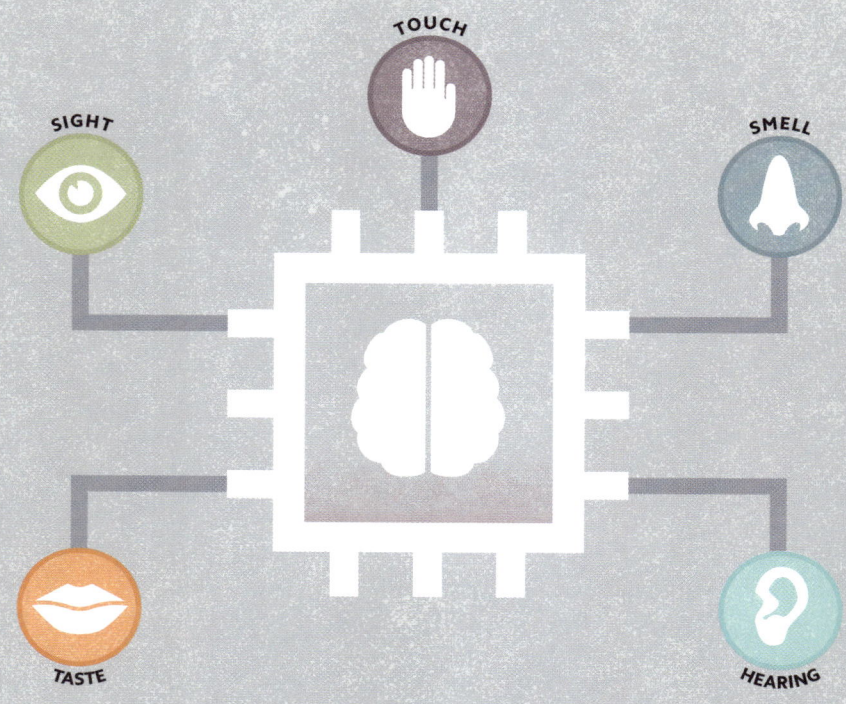

# SENSORY AI

A key aspect of human intelligence is the ability to perceive the world through sight, hearing, touch, smell, and taste. Machine perception is the ability of computers to sense their surroundings via dedicated hardware (such as cameras and microphones), and to interpret the collected data and react appropriately. This allows computers to receive information from sources other than a keyboard and a mouse, which is a step towards aligning AI with human intelligence. Machine perception, which is vital for embodied AI (see p.118), includes computer vision (see p.110), machine hearing (see opposite), machine touch, machine smelling, and machine taste.

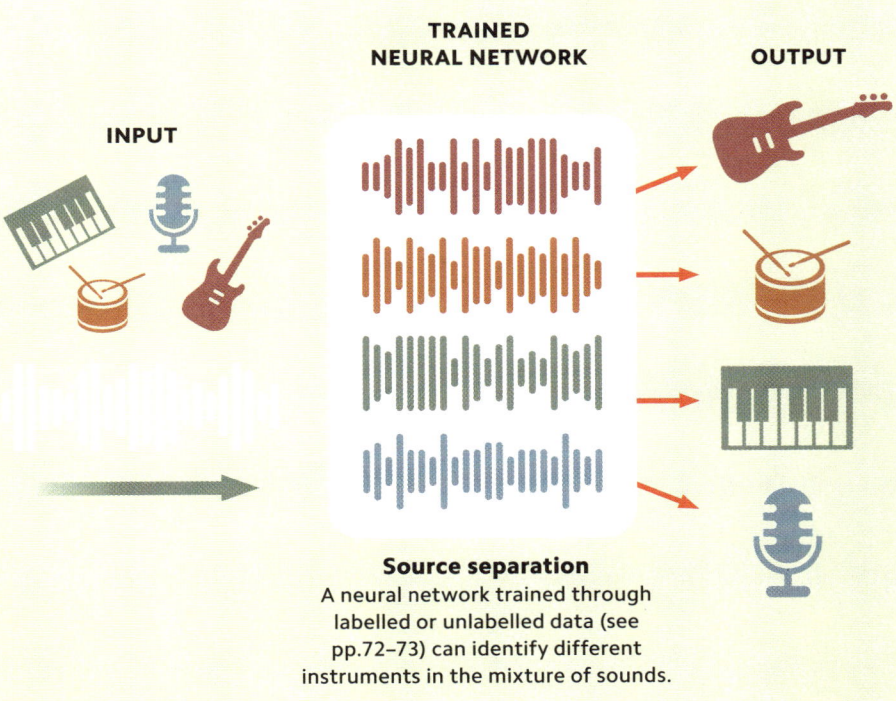

**Source separation**
A neural network trained through
labelled or unlabelled data (see
pp.72–73) can identify different
instruments in the mixture of sounds.

# PROCESSING SOUND

Machine hearing is the ability of a computer to sense and process
audio data, such as music or human speech. This interdisciplinary field
employs both classical (see p.35) and statistical (see p.57) approaches
to AI. Engineers developing machine-hearing technologies attempt to
replicate the abilities of the brain that people typically take for granted,
such as focusing on a specific sound amid background noise. Speech
recognition is a complex subfield within machine hearing. It aims to
comprehend meaning in spoken language, often using
deep learning (see p.86) to train models.

# MIMICKING SIGHT

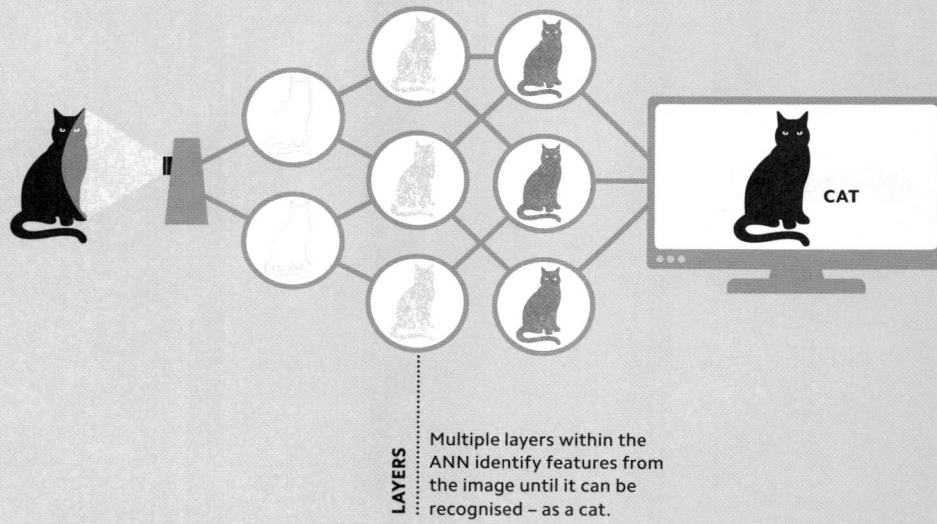

**LAYERS** Multiple layers within the ANN identify features from the image until it can be recognised – as a cat.

Computer vision is the ability of a computer to recognize images and videos – for example, to understand that a certain arrangement of pixels is associated with a picture of a cat. Engineers working in computer vision aim to automate the tasks performed by biological visual systems, such as the human eye and parts of the nervous system. The rise of deep learning (see pp.86) using multi-layered artificial neural networks (ANNs, see p.76) and the availability of very large training data sets online have greatly advanced the field. Computer vision is used in many areas, including facial recognition (see opposite).

# FACIAL RECOGNITION

Facial recognition is a form of computer vision technology
(see opposite) that matches photographs or videos of human
faces to those stored in a database. An image of a face
is captured, and its distinctive features, such as
the distance between the eyes, are mapped to
create a unique "faceprint" that is then compared
with known faceprints. Facial recognition is mainly used for
security, such as an authentication process on phones, and law
enforcement, such as to
identify someone from
a database of known
offenders.

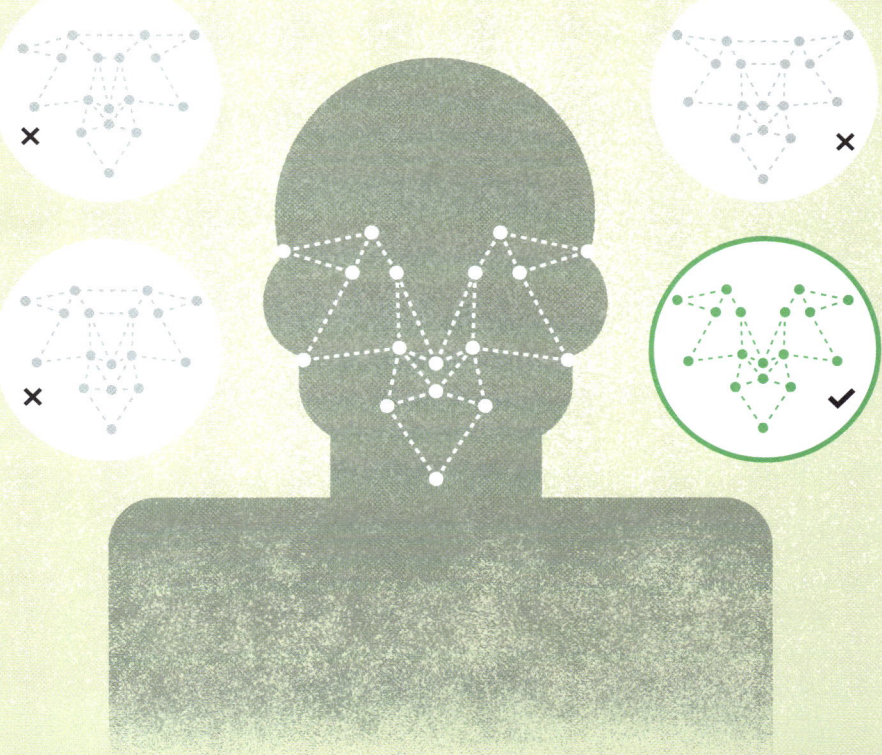

# UNDERSTANDING WORDS

The ability of computers to "understand" and generate natural language – that is, language as it is typically spoken and written by humans – is a key element of mimicking human intelligence. This idea lies at the heart of the Turing test (see pp.130–131). Natural language processing (NLP), the research field dedicated to developing this ability, brings together AI, linguistics, and other disciplines. In the 1950s, researchers tried to emulate "linguistic intelligence" by providing computers with collections of hand-written language rules. More recently, the explosion in computing power and big data (see p.33) has enabled machine learning – particularly deep learning – to be integrated into NLP with impressive results. Among its many applications, NLP is used in machine translation (see p.114) and virtual assistance (see p.116).

**Elements of NLP**
There are five elements to NLP, which involve arranging letters into words and interpreting the intended meaning of sentences.

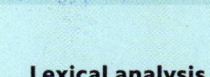

**Lexical analysis**
Lexical analysis involves structuring an example of natural language into words, sentences, and paragraphs.

**Syntactic analysis**
The application of the formal rules of grammar to natural language is known as "syntactic analysis".

**Semantic analysis**
Semantic analysis is the process of determining the literal meaning of the words in an example of natural language.

**Discourse integration**
The meanings of consecutive sentences are considered together to give context to words and phrases.

**Pragmatic analysis**
Pragmatic analysis goes beyond the literal meaning of the words and attempts to interpret their intended meaning.

# AI INTERPRETERS

Machine translation (MT) is the use of AI in the automated translation of text or speech from one language to another. Translation is a far more complex and subtle matter than simply substituting each word for its equivalent in another language. Consequently, MT is currently used more as a tool rather than as a replacement for human translators. There are three broad approaches: "rule-based MT" relies on linguistic rules, such as grammar and syntax; "statistical MT" uses the known relationships between words to predict whole sentences and phrases; "neural MT" uses artificial neural networks (ANNs, see p.76) trained to understand languages almost as well as people do.

## MACHINE TRANSLATION IN ACTION

**Rule-based MT**
This approach gives a quick but basic translation. Text and speech can be understood, but often requires further editing.

**Statistical MT**
This approach predicts words and sentences, and may not be fully accurate. The translated text often still requires further editing.

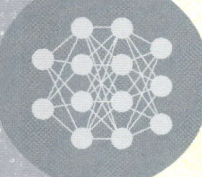

**Neural MT**
A trained ANN is accurate and can be constantly improved. Training an ANN requires huge amounts of data and is very costly.

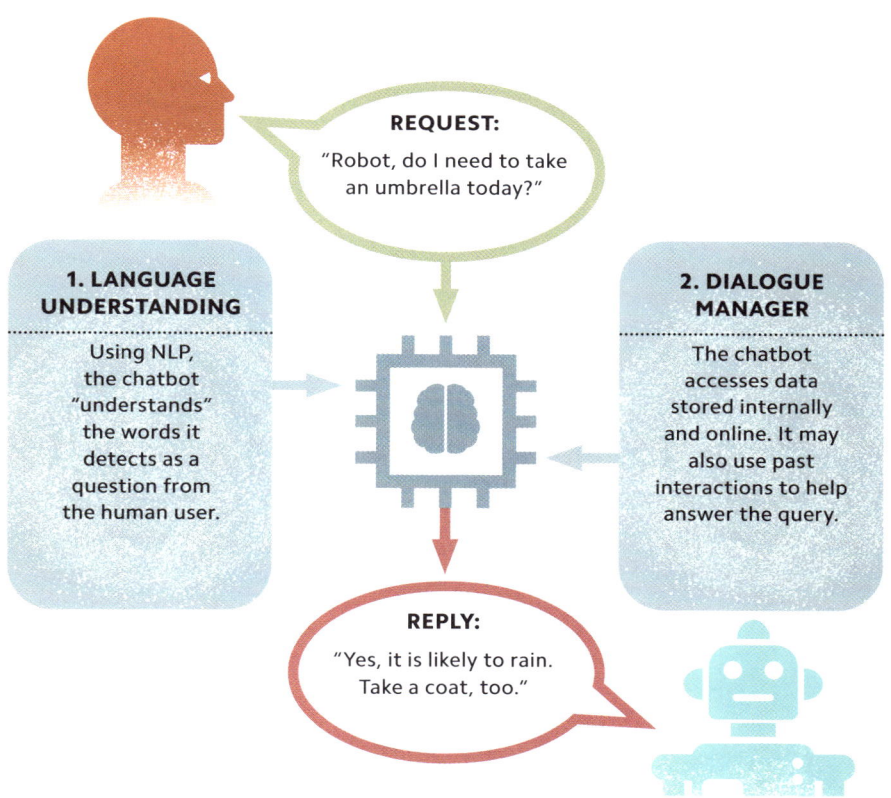

**REQUEST:**
"Robot, do I need to take an umbrella today?"

**1. LANGUAGE UNDERSTANDING**

Using NLP, the chatbot "understands" the words it detects as a question from the human user.

**2. DIALOGUE MANAGER**

The chatbot accesses data stored internally and online. It may also use past interactions to help answer the query.

**REPLY:**
"Yes, it is likely to rain. Take a coat, too."

# TALKING WITH AI

Chatbots, such those employed by virtual assistants (see p.116), are programs that can carry out conversations via text or text-to-speech. Natural language processing (NLP, see pp.112–113) helps chatbots to mimic how humans talk during conversations. Based on classical AI, statistical AI, or a combination of the two (see pp.54–55), chatbots can range in sophistication. Businesses often use basic chatbots to answer simple customer queries instead of providing immediate contact with a human employee. The most sophisticated chatbots, such as ELIZA in the 1960s, can give the impression of intelligence.

# AI HELPERS

A virtual assistant is a software application or device that uses machine hearing (see p.109) and natural language processing (NLP, see pp.112–113) to perform tasks on command, such as searching the internet, playing music, or setting timers and alerts. Basic virtual assistants are essentially chatbots (see p.115), while more complex models can interact with other "smart" devices, via the Internet of Things (see p.104), to activate systems such as domestic lighting and heating. Many virtual assistants are cloud-based, and continually use voice data for training, which enables them to get better at predicting a user's needs and preferences.

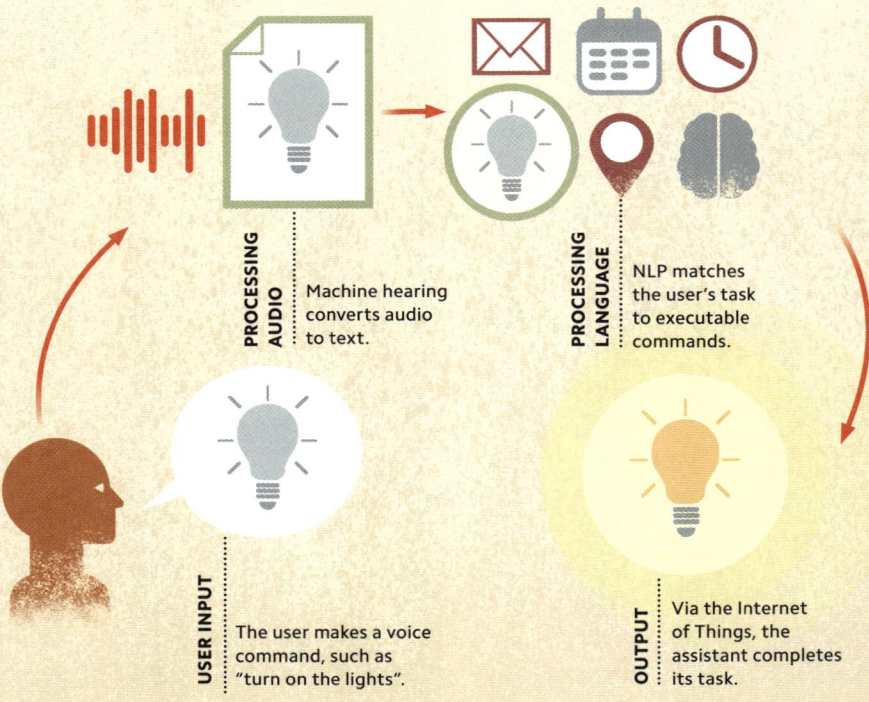

**PROCESSING AUDIO**
Machine hearing converts audio to text.

**PROCESSING LANGUAGE**
NLP matches the user's task to executable commands.

**USER INPUT**
The user makes a voice command, such as "turn on the lights".

**OUTPUT**
Via the Internet of Things, the assistant completes its task.

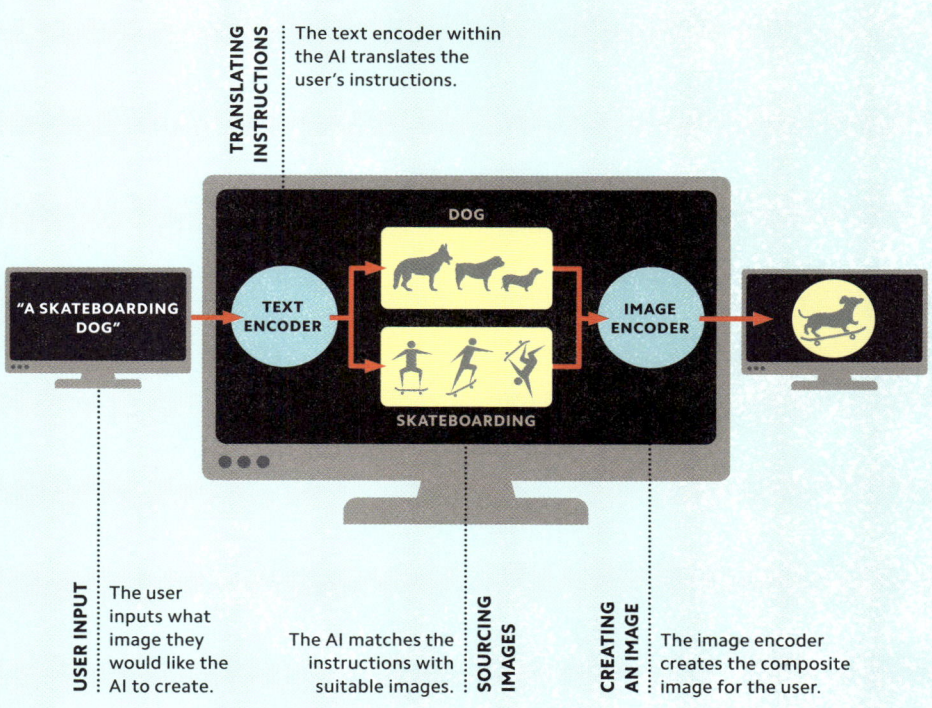

The text encoder within the AI translates the user's instructions.

"A SKATEBOARDING DOG"

TEXT ENCODER

DOG

SKATEBOARDING

IMAGE ENCODER

USER INPUT
The user inputs what image they would like the AI to create.

The AI matches the instructions with suitable images.

SOURCING IMAGES

CREATING AN IMAGE

The image encoder creates the composite image for the user.

# AI ARTISTS

Generative AI is the field dedicated to synthesizing new content, such as images, audio, text, or video, based on an input in any of these formats. For instance, a generative AI model could be trained to produce an image of a cartoon giraffe when prompted with the text input "cartoon giraffe". AI image-generation has existed since the 1960s, and can use a variety of classical and statistical techniques. Recently, however, generative adversarial networks (GANs, see p.87) have proved such effective "artists" that they have prompted debate about whether art can be considered uniquely human.

# INTELLIGENT ROBOTS

AIs that are designed to interact physically with their environments are known as "embodied AIs". Such AIs, which include robots, mimic not only human cognitive intelligence, but human physical behaviour as well. They do so with the help of sensors, motors, and other hardware, which enable them to perceive (see p.108), move in (see p.120), and affect (see p.121) their three-dimensional environments. Constructing such machines is an important step forwards in AI, since much of what we consider to be intelligence in human beings involves our ability to interact with our surroundings. Embodied AIs already include robotic vacuum cleaners and lawn mowers.

# AI COMPANIONS

A social robot is an embodied AI (see opposite) that is capable of interacting socially with humans, using speech, movement, facial expressions, and other human-like behaviours. Social robots are limited as companions, since it is difficult to replicate many basic human abilities, such as manipulating objects (see pp.52–53) or understanding tone of voice. While largely treated as a novelty, they are nevertheless sometimes used in health and social care to alleviate loneliness, depression, and anxiety. Although they can come in any shape and size, most social robots are humanoids.

# MOVEMENT AND MOBILITY

Many robots are kept in stationary positions, such as on production lines, but some, such as drones, can move around and explore their environments. These mobile robots have varying degrees of autonomy: some are controlled remotely by human beings, while others can navigate without human intervention. A fully autonomous robot has an AI that can process data collected by sensors, such as optical cameras and LIDAR (see p.122), to plan the path ahead.

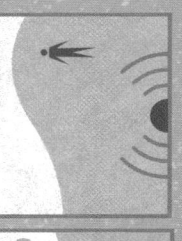

**LEARNING**
A feedback loop improves the accuracy of the AI.

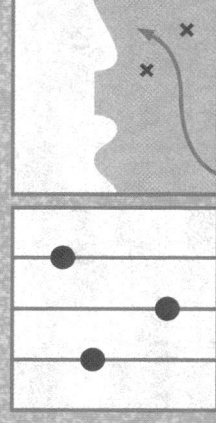

**SENSING**
Sensors, such as cameras, pick up information about the environment.

**FUSION**
An AI organises the information into a model of the environment.

**PERCEPTION**
The AI identifies its own location and its destination with the sensory input.

**PLANNING**
After studying the model, the AI plots the best path to its destination.

**CONTROL**
The AI steers the robot around objects to reach its destination.

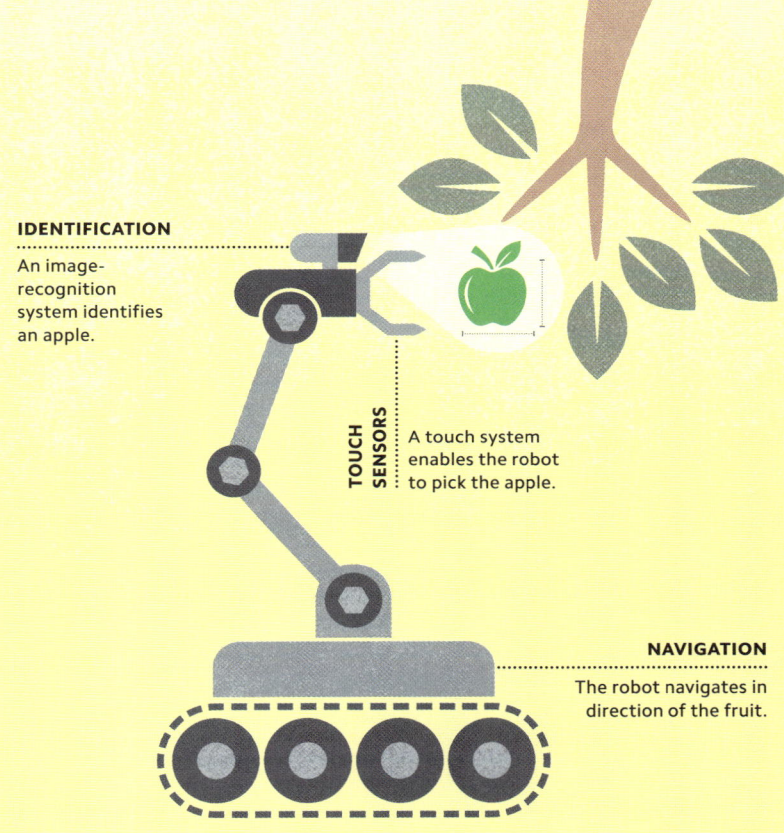

**IDENTIFICATION**

An image-recognition system identifies an apple.

**TOUCH SENSORS**

A touch system enables the robot to pick the apple.

**NAVIGATION**

The robot navigates in direction of the fruit.

# MANUAL DEXTERITY

One of the greatest challenges in robotics is building machines that can interact physically with their environments. To perform even the simplest human action, such as picking an apple, a robot must have an excellent sense of sight, as well as a sense of touch, which enables it to apply just the right amount of pressure to manipulate objects correctly. Many such robots are currently used in controlled settings, such as factories, but they may soon be sophisticated enough to help with domestic chores in people's homes.

# DRIVERLESS CARS

Autonomous vehicles are examples of mobile robots (see p.120). They use systems incorporating sensors, AI, and actuators (see p.27) to assist or wholly replace the human operator of a vehicle, whether on land or sea, or in the air. "Self-driving" or "driverless" cars are a category of autonomous vehicle under development, and cars incorporating semi-autonomous technology are now available. Their arrival on roads is raising complex legal and ethical questions, such as who would be responsible for accidents caused by AI-controlled cars (see p.152).

**RADAR** Radar detects other vehicles and reveals their speed, distance, and direction of travel.

**LIDAR** Laser scanning produces a 3D map of the vehicle's surroundings.

**CAMERA** A camera reads road signs and identifies traffic-light colours.

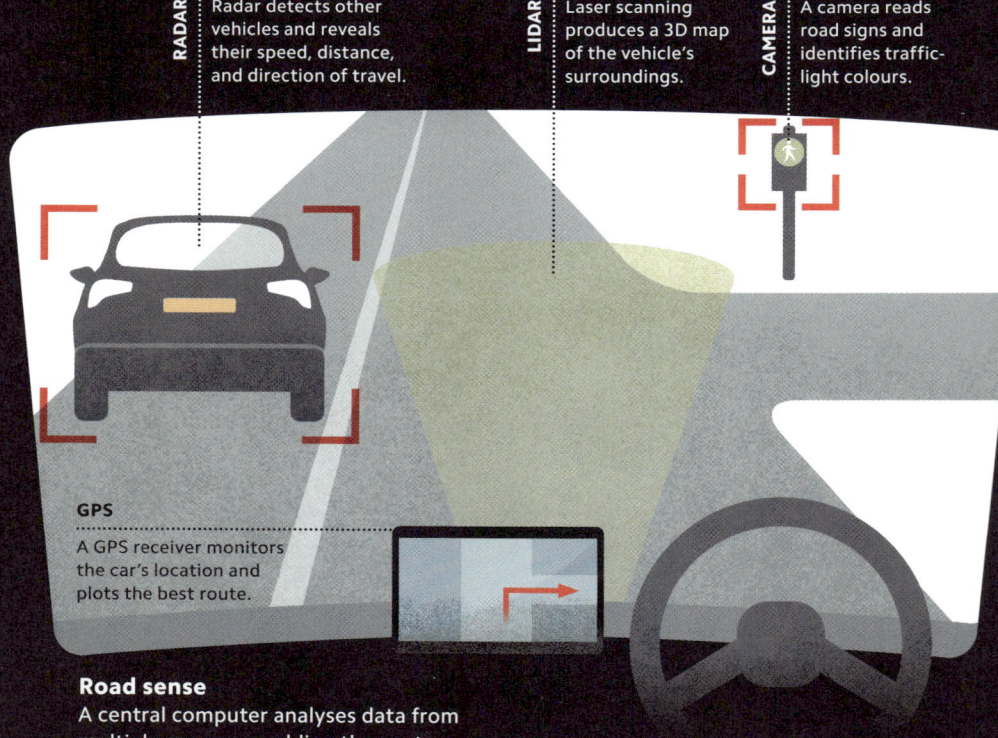

**GPS**
A GPS receiver monitors the car's location and plots the best route.

**Road sense**
A central computer analyses data from multiple sensors, enabling the car to "understand" the driving environment.

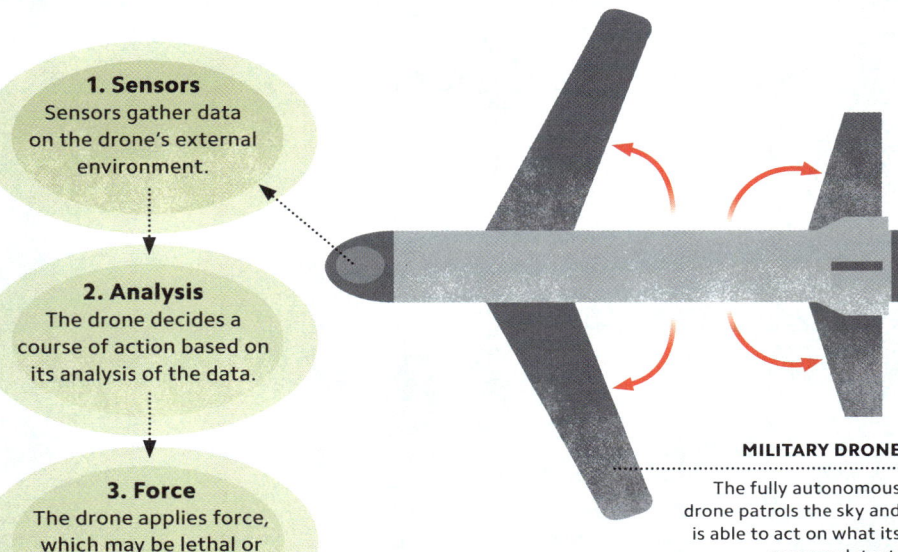

**1. Sensors**
Sensors gather data on the drone's external environment.

**2. Analysis**
The drone decides a course of action based on its analysis of the data.

**3. Force**
The drone applies force, which may be lethal or non-lethal.

**MILITARY DRONE**
The fully autonomous drone patrols the sky and is able to act on what its sensors detect.

# AI AND WARFARE

Military needs drive much AI innovation. This has led to the creation of sophisticated autonomous systems that can perform military tasks with little or no human intervention. Some, including reconnaissance drones, are non-lethal. Others, such as sentry guns, are deadly weapons in their own right, capable of identifying, locking onto, and firing at targets. There is much debate over whether to ban the deployment of lethal weapons that are fully autonomous – those that enable rapid response by removing the need for a human to give the final order to attack.

# PHILOSO

# ARTIFIC

# INTELLI

# PHY OF
# IAL
# GENCE

**AIs are designed** to mimic human behaviour – to calculate
the way we do, or, in the case of androids, to interact with the
environment with human-like agility. However, as AIs become
ever more sophisticated, the question arises as to where we
should draw the line between the human and the artificial.
Or, to put the question another way: at what point should we
say that an AI is, in fact, a person – has all of the qualities that
a human has, and so should be granted rights? The philosophy
of AI addresses this central question. It examines the concepts
of free will and consciousness, and asks what the difference is
between an intelligence that has evolved biologically and one
that has been built by human beings.

# HUMAN-LIKE AI

For many scientists, building an AGI (artificial general intelligence) is the ultimate goal of AI research – although it may never be achieved. An AGI would be as intelligent as a human being, and may even have other human faculties, such as emotions or even consciousness. Another name for AGI is "strong AI", a name that contrasts it with "weak AI", which refers to all other AIs that are built to perform specific tasks. Unlike a weak AI, an AGI would have something like intuition – the ability to know that something is true without resorting to conscious reasoning.

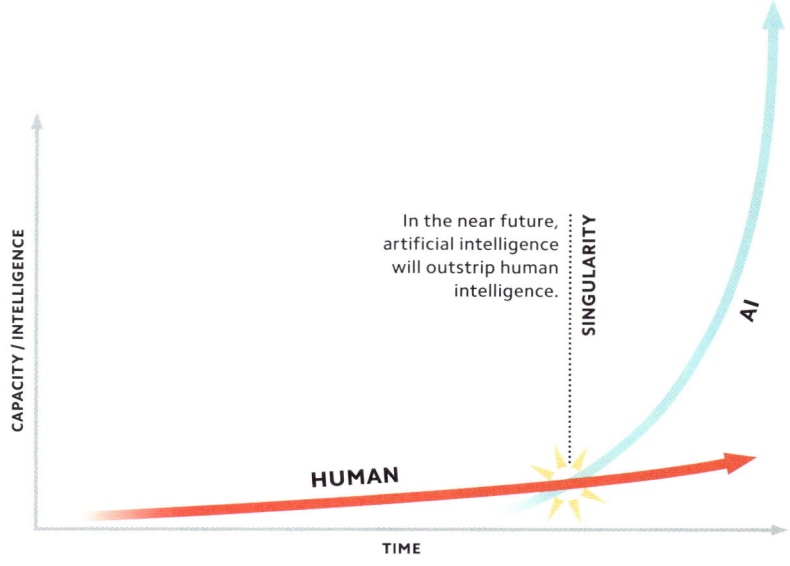

CAPACITY / INTELLIGENCE

In the near future, artificial intelligence will outstrip human intelligence.

SINGULARITY

AI

HUMAN

TIME

# THE POINT OF NO RETURN

In cosmology, a "singularity" is a point in space at which the familiar laws of physics break down, creating a phenomenon known as a "black hole". In AI, the singularity is the name given to the point in time at which a machine will become as smart as the people who built it, and therefore clever enough to improve itself. Such a machine would be able to operate at the lightning speeds of a supercomputer, and so would swiftly achieve incredible abilities – including the ability to design AIs itself. The singularity could therefore transform the world in ways that we simply cannot predict.

# WHERE IS CONSCIOUSNESS?

For centuries, philosophers have debated the question of how the mind and the brain interact – or, more broadly, how such a thing as consciousness can even exist in a physical world. The debate intensified in the 17th century, when scientists proposed that the universe is like a machine – a clockwork mechanism whose workings are in principle predictable. However, German philosopher Gottfried Leibniz (1646–1716) argued that if the physical world is mechanical, then the human brain must be linked to the rest of the body by the biological equivalents of cogs and pulleys. But if that is the case, he argued, then there is no place in the brain for consciousness, which he believed cannot be explained mechanically.

**Machine**
According to Leibniz, no physical structure, such as an AI, could be conscious because everything about it can be explained in physical terms.

**Human**
Leibniz's argument also applies to the human brain; since the brain is entirely physical, then consciousness is irrelevant to how it works.

Today many scientists argue that debates about how the mind interacts with the body (see opposite) are futile, and that the mind is simply the brain in action – the equivalent of software running on the hardware of the brain. This approach, known as "functionalism", was summed up by Dutch computer scientist Edsger Dijkstra (1930–2002), who said: "The question of whether computers can think is like the question of whether submarines can swim". In other words, whether or not we say that an AI can "think" or be "conscious" is simply a matter of linguistic convention, not one of scientific discovery. Functionalists focus on what things do rather than what they are – and, they argue, if we want to say that submarines "swim", then they swim.

# DO SUBMARINES SWIM?

# THE IMITATION GAME

Alan Turing (see pp.18–19) devised a test, now called the Turing test, that provides a means for judging whether or not a machine is intelligent. The test is based on a Victorian parlour game, in which one person tries to work out whether another person, who is hidden behind a screen, is male or female, judging by the answers they give to certain questions. In the Turing test, both a human and a computer are hidden behind a screen, and an examiner supplies them with mathematical problems to solve. If both sets of answers are correct, then the examiner cannot say which are the computer's and which are the human's. The computer has therefore passed the test, and can be said to be intelligent.

**HUMAN** The human provides the examiner with printed responses to questions.

**COMPUTER** The computer provides the examiner with printed responses to the same questions.

The printed answers from the computer and the human are compared by the examiner.

"If a machine is expected to be infallible, it cannot also be intelligent."
Alan Turing

**EXAMINER**

Can the examiner correctly identify which responses are from the computer and which are from the human?

# INTELLIGENCE METRICS

The Turing test is the best-known test of AI (see p.130–131), but it is not the only one. The "coffee test" for AI robots asks if an AI robot placed into a random person's home could make a cup of coffee. The "flatpack test" asks whether an AI robot could put together an item of furniture without help. Finally, the "employment test" asks if a human-level AI robot could replace a human in a particular occupation (see p.146).

> "Machine intelligence is the last invention that humanity will ever need to make."
> **Nick Bostrom**

# MACHINES AND UNDERSTANDING

American philosopher John Searle (1932–) rebutted the idea that machines can think by arguing that while machines follow rules, they are incapable of understanding them (see pp.130–131). In what he called the "Chinese Room" thought experiment, Searle imagined a person in a room receiving questions written in Chinese. If the person had the appropriate rule book, they would be able to reply to the questions in writing, without actually understanding either the questions or the answers. Searle argued that to say that a computer can think is similar to saying that the person in the example understands Chinese.

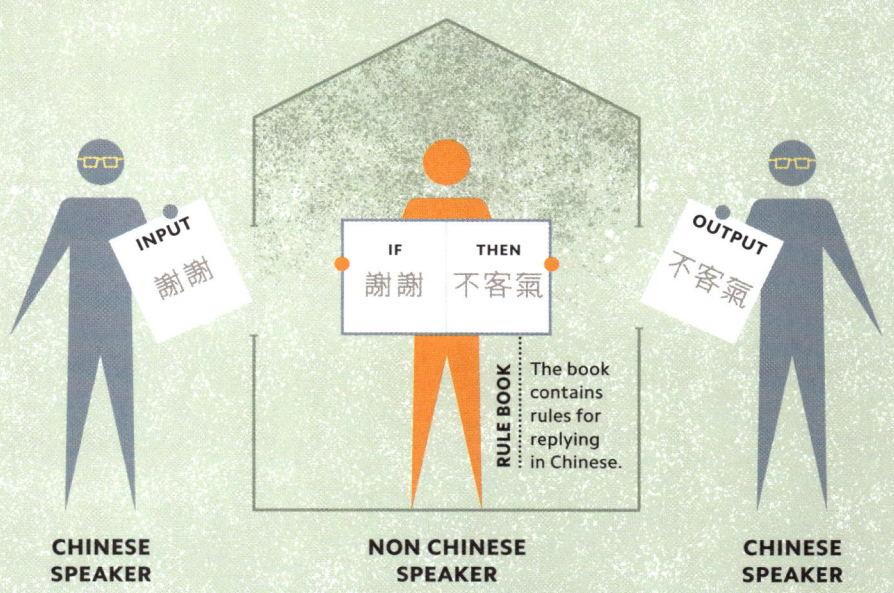

INPUT

謝謝

IF THEN

謝謝 不客氣

OUTPUT

不客氣

RULE BOOK

The book contains rules for replying in Chinese.

**CHINESE SPEAKER**

**NON CHINESE SPEAKER**

**CHINESE SPEAKER**

# PHILOSOPHICAL ZOMBIES

Many philosophers question whether or not an AI could ever be conscious (see pp.128–129), or be alive in the way that organisms are. Some claim that such developments are impossible, because AIs are entirely mechanical, and are designed specifically to mimic human behaviour. If this is true, then even the most lifelike AIs (see p.126) would be like zombies: they would have no "inner life", and could only ever simulate having emotions, interests, preferences, or opinions.

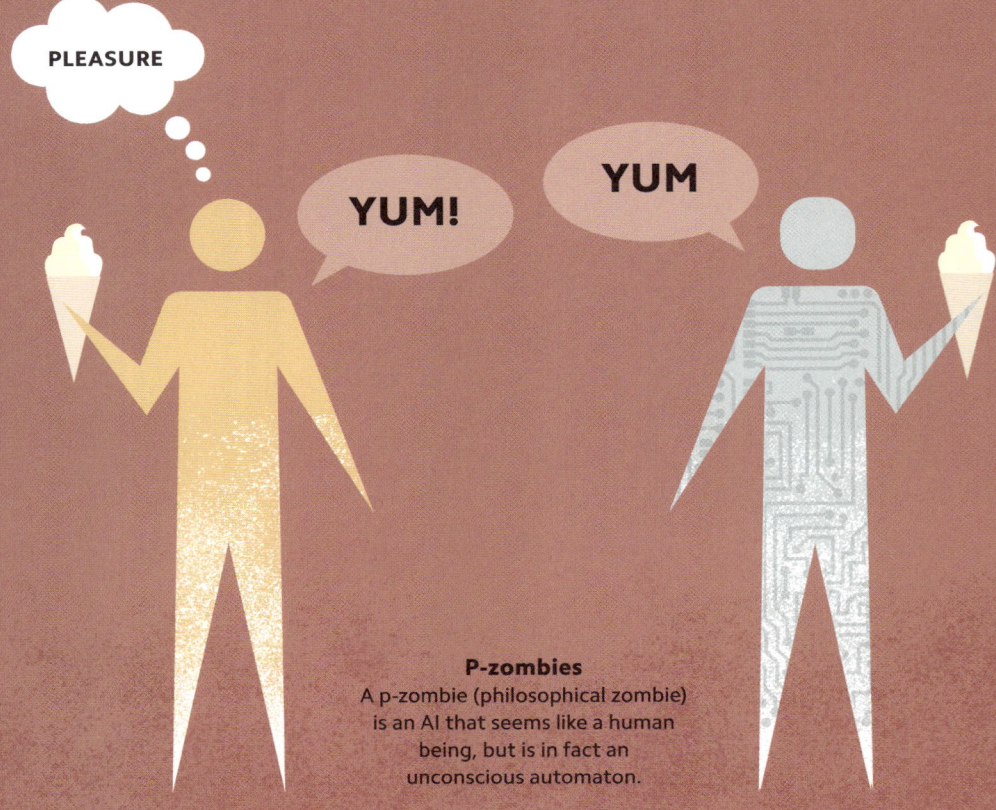

**P-zombies**
A p-zombie (philosophical zombie) is an AI that seems like a human being, but is in fact an unconscious automaton.

# A NEW KIND OF PERSON

Many scientists argue that, one day, AIs will be so lifelike that they should be treated like human beings. They claim that since humans have rights on the basis that they have free will, AIs that pass a "free will test" should therefore have the same protections under law. This means that, in the future, an AI could claim ownership of its intellectual property, and even be penalized for making mistakes. Legally, such an AI would no longer be a machine, but a person – effectively, a new kind of human being.

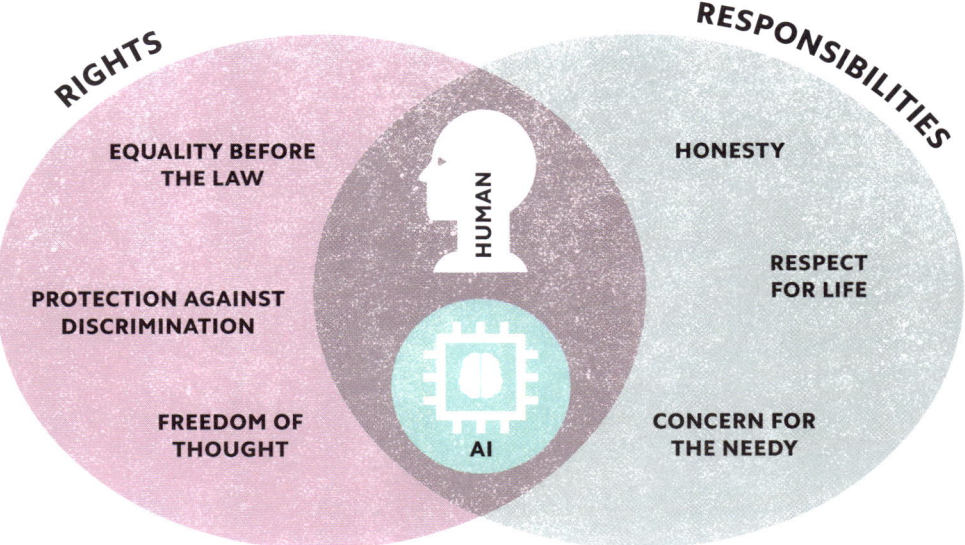

**RIGHTS**

**EQUALITY BEFORE THE LAW**

**PROTECTION AGAINST DISCRIMINATION**

**FREEDOM OF THOUGHT**

HUMAN

AI

**RESPONSIBILITIES**

**HONESTY**

**RESPECT FOR LIFE**

**CONCERN FOR THE NEEDY**

**Personhood**
If an AI is treated like a person, it may be granted both rights and responsibilities.

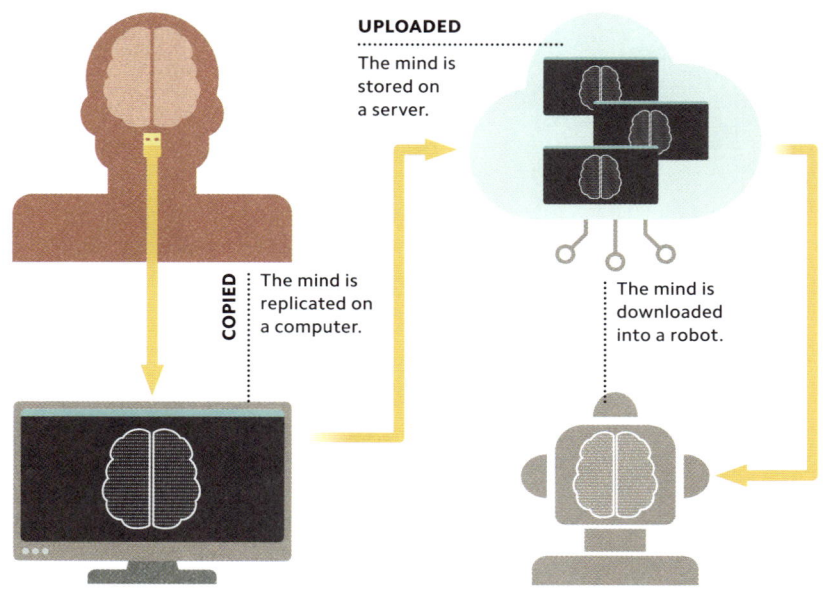

**UPLOADED**
The mind is
stored on
a server.

**COPIED**
The mind is
replicated on
a computer.

The mind is
downloaded
into a robot.

# REPLICATING
# THE MIND

According to the principle of multiple realizability (see p.20),
the same computer programs can be run, or "realized", on
different devices. Computationalists (see p.12) argue that
human thought is computable, and so can be realized by a
machine as well as a brain. If this is true, then it should be
possible to write a program that replicates a human mind,
which could then be copied and transferred like any other
program. This means that a person's mind could be
uploaded to a remote server, downloaded to a robot,
and even duplicated innumerable times.

The thoughts of others are closed off to us.

**Human–human interaction**
An individual knows what they themselves mean by "beetle", but they cannot be sure that it means the same thing to someone else.

# TRANSPARENT THINKING

The philosopher Ludwig Wittgenstein (1889–1951) argued that a person's thoughts were like objects in a closed box – a box into which only they could "see". We can never know what another person is really thinking or exactly what things mean to them, since the "box" is closed to us. A machine intelligence, however, could be examined in ways a human mind cannot. If the machine said it was thinking of a beetle, its programming could be exposed – "opening the box" – to show precisely what it meant by "beetle". Such developments might, in turn, shed light on the mechanisms of human consciousness and thinking.

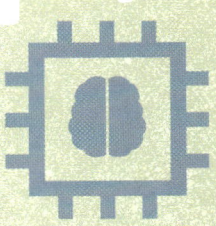

**Human–AI interaction**
A person thinking of a certain beetle might be able to examine an AI's programming to see if it is thinking of the same beetle.

LIVING

ARTIFIC

INTELLI

# W I T H

# I A L

# G E N C E

**Like the combustion engine** and the internet, AI is a general-purpose technology that is changing how we live our lives. There is no doubt that AI is here to stay. The only question is: how do we adapt to it? Society is already grappling with AI-driven technological unemployment, algorithmic biases that worsen inequality, and an entirely new kind of conflict: cyberwarfare. Some researchers even claim that AI is a threat to our species. However, AIs also make life incalculably better for us, particularly in the fields of medicine, finance, and agriculture. How to ensure that AIs are only ever used for good remains an urgent and unresolved subject of debate.

# MYTH OR REALITY?

The term "AI" is sometimes used to make exaggerated claims about the potential threats or benefits of machine learning (see pp.58–59). Some of these AI myths evoke fear, predicting killer robots, rogue algorithms, and other existential risks (see p.154). Others inflate the powers of machine learning, claiming that AI has "agency" – able to think for itself – and that it is objective, efficient, and powerful. In reality, AI applications have, at best, limited and specific abilities, and cannot think for themselves. They are only capable of what they are programmed to do.

AI will become more capable over time, although it will only pose a risk to humanity if humans make that possible. The only true threats that exist are the biases, intentions, and limitations of its programmers, and the data used to train it (see pp.142–145).

> "The real risk with AI isn't
> malice but competence."
> Stephen Hawking

## Behind the scenes

However AIs are portrayed, whether
as threats or saviours of humanity,
they are not autonomous. AI is
controlled by people.

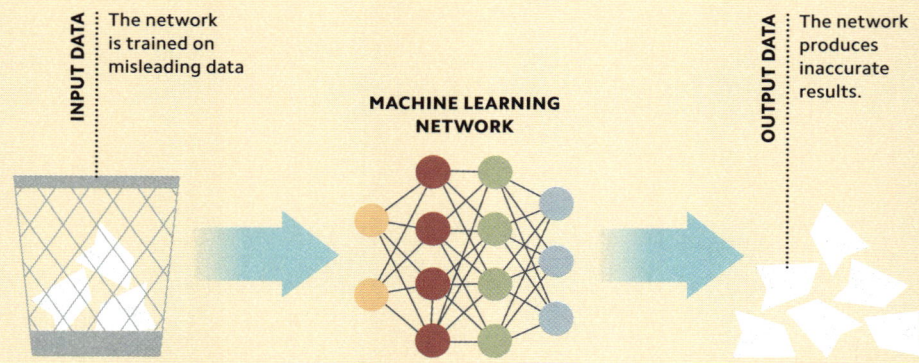

INPUT DATA

The network is trained on misleading data

**MACHINE LEARNING NETWORK**

OUTPUT DATA

The network produces inaccurate results.

# GARBAGE IN, GARBAGE OUT

Machine-learning networks (see pp.58–59) are only as good as the data on which they are trained. The most common cause of inaccurate results from an AI system is poor-quality training data, which includes input data that is incomplete, poorly labelled, full of errors, or biased (see opposite). For example, predictive AI systems (see pp.70–71) trained on inconsistent and incorrect historical data will produce useless predictions. In the field of computer science, the idea that bad inputs produce bad outputs is informally summarized as "garbage in, garbage out", or "GIGO".

# PREJUDICED OUTCOMES

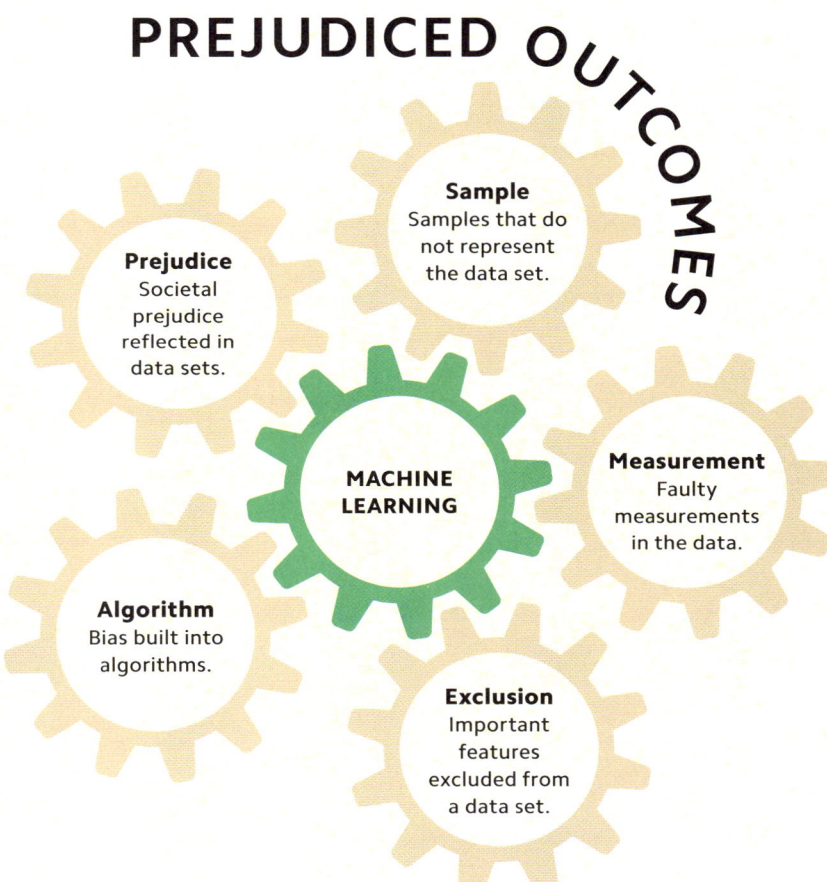

**Prejudice**
Societal prejudice reflected in data sets.

**Sample**
Samples that do not represent the data set.

**MACHINE LEARNING**

**Measurement**
Faulty measurements in the data.

**Algorithm**
Bias built into algorithms.

**Exclusion**
Important features excluded from a data set.

The term "AI bias" is used to describe AI systems that produce unfair results for particular groups of people. AI biases often reflect prejudices in society about gender, ethnicity, culture, age, and many others. Bias usually stems from the programmers themselves, via their algorithms and their interpretations of results, and from the data sets used to train AIs (see opposite). To combat this, programmers test their models to ensure that societal bias is not reflected in their results and use data sets that are representative.

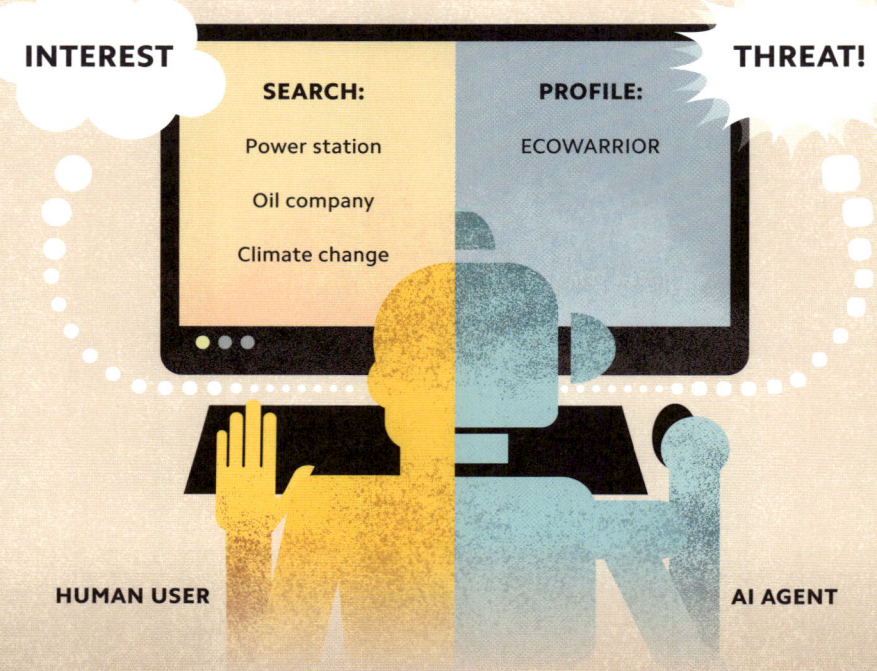

INTEREST

THREAT!

SEARCH:

Power station

Oil company

Climate change

PROFILE:

ECOWARRIOR

HUMAN USER

AI AGENT

# MAKING ASSUMPTIONS

Using personal data to try to predict an individual's future desires, opinions, and activities is known as "profiling". In AI, machine-learning tools can be trained on large data sets to become expert at predicting, for example, the kind of internet content a user might like to see based on their viewing history. However, profiling can be problematic, since it can lead to false and even damaging predictions due to biases built into data sets and algorithms (see p.143). In order to root out such biases, it is essential that AI decision-making processes are transparent (see opposite).

# TRANSPARENT PROCESSING

Machine-learning models process data and make predictions using highly complex artificial neural networks (ANNs, see p.76). The inner workings of these models are often said to be a "black box" since they are too complicated and abstract for humans to "observe". This means that the results they produce cannot be properly understood and checked for errors or biases. An alternative approach, known as "interpretable machine learning", or "white box AI", shines a light into the black box. White box AIs are designed to give not just the result, but a breakdown of the processes they followed to reach it.

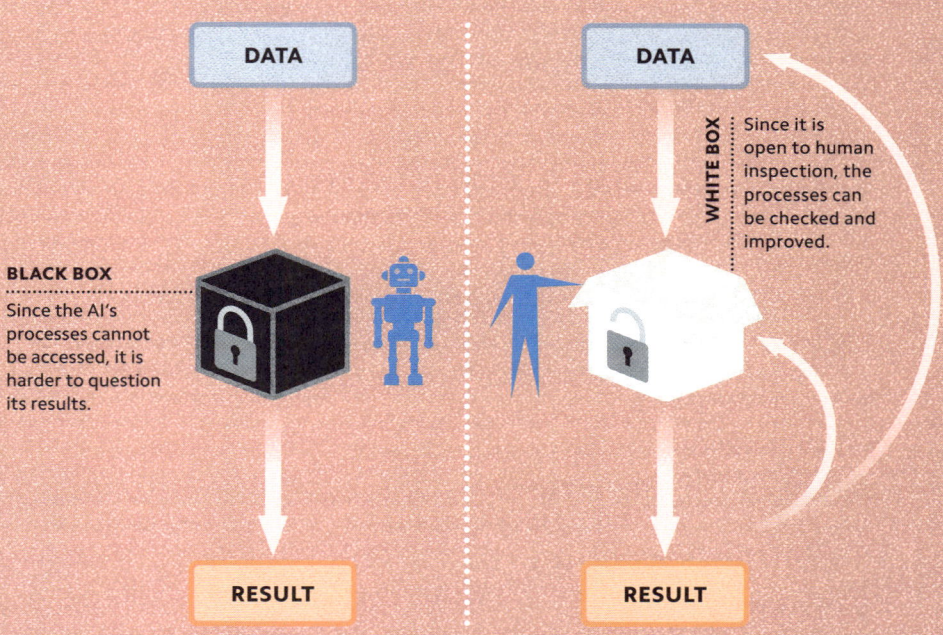

**DATA**

**DATA**

**WHITE BOX**

Since it is open to human inspection, the processes can be checked and improved.

**BLACK BOX**

Since the AI's processes cannot be accessed, it is harder to question its results.

**RESULT**

**RESULT**

# AN AI
# WORKFORCE

The replacement of human beings by machines in the
workforce is known as "technological unemployment". Up until
now, this phenomenon has not led to mass unemployment, because
machines greatly increase productivity, which in turn stimulates the
economy and creates new job opportunities. However, if AIs begin
to pass the employment test (see p.132), and achieve the intelligence
level of AGIs (see p.126), then, one day, there may be few jobs left
for human beings to do. Under such circumstances, the challenge for
governments will be how to support the masses of unemployed
people, which may include providing universal income – a regular
payment to each member of society.

# THE AI BALANCE

**Balance of power**
The democratic approach to AI aims to ensure that the technology benefits everyone, rather than a rich, powerful elite.

**THE WEALTHY**

**THE MASSES**

AI has the potential to increase productivity and generate income and opportunity. Shared by all, these benefits could create a more equal world, but if concentrated in the hands of the wealthy and powerful, the gulf between rich and poor will widen. Bias in design, data, and how and where AIs are used can exacerbate social divides, increase inequality, and lead to hazardous and discriminatory applications. Attempts to mitigate these risks include inclusive design and embedding AIs with values, such as fairness and accountability.

Different shades of opinion are excluded.

# AN ECHO CHAMBER

AI algorithms are increasingly used to curate the content people see online – for example, on social media. An unintended consequence of this has been the creation of "filter bubbles", whereby people are only shown content that tallies with or amplifies their own opinions, while alternative views get filtered out. This occurs due to "recommendation algorithms" (see p.95) repeatedly showing users material similar to what they have viewed in the past – encouraging biased thinking.

ALLOWED IN

Compatible opinions are the only ones that make it through.

Some people fear that, in the future, AIs may attempt to take control of their own actions.

NO CONTROL

# THE LIMITS OF CONTROL

The rogue AIs of dystopian science fiction are imaginary, but at their root lies a serious issue: the problem of control. If an AI is to maximize its usefulness, it will need to be autonomous to an extent – that is, capable of independent decision-making. However, the more autonomous and powerful an AI becomes, the harder it will be to control. A fully autonomous AI might be able to ignore or contradict the instructions of its controllers, and even take active steps to maintain its independence. Once an AI is beyond human influence and restraint, its behaviour would be unpredictable.

# RIGHT VS WRONG

As AIs become ever more intelligent, the question of how to ensure that they behave ethically becomes increasingly important (see opposite). Machine-learning tools have neither agency nor values, and so cannot be relied upon to offer suggestions that are in the best interests of humanity, or do not favour one social group over another. The only way to ensure that AIs think ethically is to program them with ethical principles, although then the question becomes: whose ethics? Ideally, an AI should have equal respect for all humans, and be able to detect and compensate for bias.

|  | **UNETHICAL DESIGN** | **ETHICAL DESIGN** |  |
|---|---|---|---|

**Black box**
Decision-making is not transparent. People cannot see why the AI has made the decision it has.

**White box**
Decision-making is transparent. How an AI makes its decisions can be seen and judged.

**Privacy violations**
Individuals are not in control of their data; they do not know who can see it or how it is being used.

**Privacy protections**
Personal data is kept private; the individual retains control over who can see it and how it is used.

**Algorithm bias**
Bias is designed into the AI, and those who control it have the most power.

**Algorithm fairness**
Bias is designed out of the AI at every stage, from data-collection to final application.

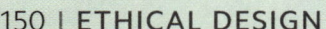

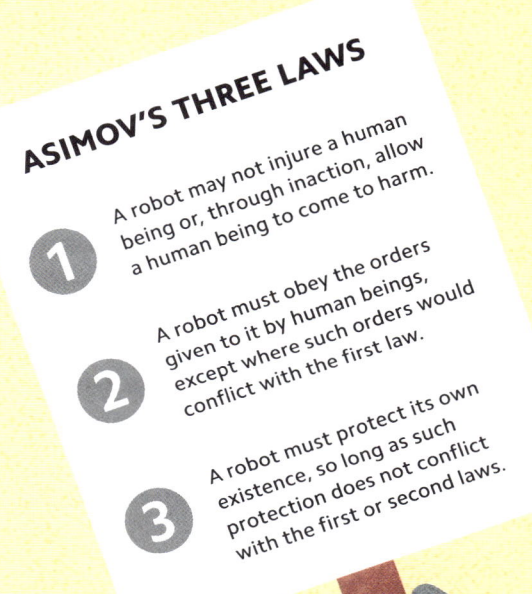

**ASIMOV'S THREE LAWS**

1. A robot may not injure a human being or, through inaction, allow a human being to come to harm.

2. A robot must obey the orders given to it by human beings, except where such orders would conflict with the first law.

3. A robot must protect its own existence, so long as such protection does not conflict with the first or second laws.

# IN-BUILT ETHICS

One way of ensuring that AIs behave ethically (see opposite) is to program them with specific ethical rules or laws – a process known as "terminal value loading". The classic illustration of this can be found in the science fiction stories of Isaac Asimov (1920–92), who formulated what he called the "three laws of robotics" (see above). However, as his stories explore, terminal value loading is far from foolproof, since even the simplest laws can generate contradictions. For example, an AI may be instructed not to harm a human being, but doing so may be the only way of saving a person's life.

# WHO IS TO BLAME?

Some researchers argue that, one day, AIs will not only be as intelligent as human beings, they will also have human-like personalities, and so should be granted human rights (see p.135). If an AI is given such rights, lawmakers would have to decide where to draw the line between holding the AI or its makers responsible for its actions. If the AI is deemed culpable for breaking the law – in other words, that it acted on its own free will – then it would have to suffer the appropriate sanctions or punishments for its actions. Like a human being, it could also be required to make amends for what it has done, and be open to reforming its character.

**WHAT SHOULD WE ALLOW?**

UNACCEPTABLE RISK

**PROHIBITED**
AIs that could freely cause harm if left unregulated.

HIGH RISK

**Highly regulated**
AIs involved with safety, law, employment, and education.

MINIMAL RISK

**Partially regulated**
AIs that can interact with humans, understand emotions, and recognize faces.

LOW RISK

**Unregulated**
Everyday AIs, such as AI-enabled computer games and spam filters.

Concerns about the dangers that AIs may pose in the future have fuelled calls for AI research to be regulated. However, many scientists argue that regulating research will stifle innovation, and give unregulated countries a dangerous advantage. A compromise, proposed by European regulators, is to scale regulation according to risk. Low-risk applications of AI should have little or no regulation; high-risk applications should be controlled; and the most risky applications should be forbidden.

# EXISTENTIAL RISKS

One possible threat posed by AI is known as the "alignment problem", whereby the goals and values of an AI do not align with those of humanity. Named after a scene in the Disney cartoon *Fantasia*, in which a sorcerer's apprentice makes a broom multiply uncontrollably, "Sorcerer's Apprentice Syndrome" neatly illustrates the problem in the form of a thought experiment. An AI is given the task of optimizing the production of paperclips, but believes that its job is only done when it has converted the entire planet into paperclips. It does so because it does not realize that it must prioritize human life over paperclip production.

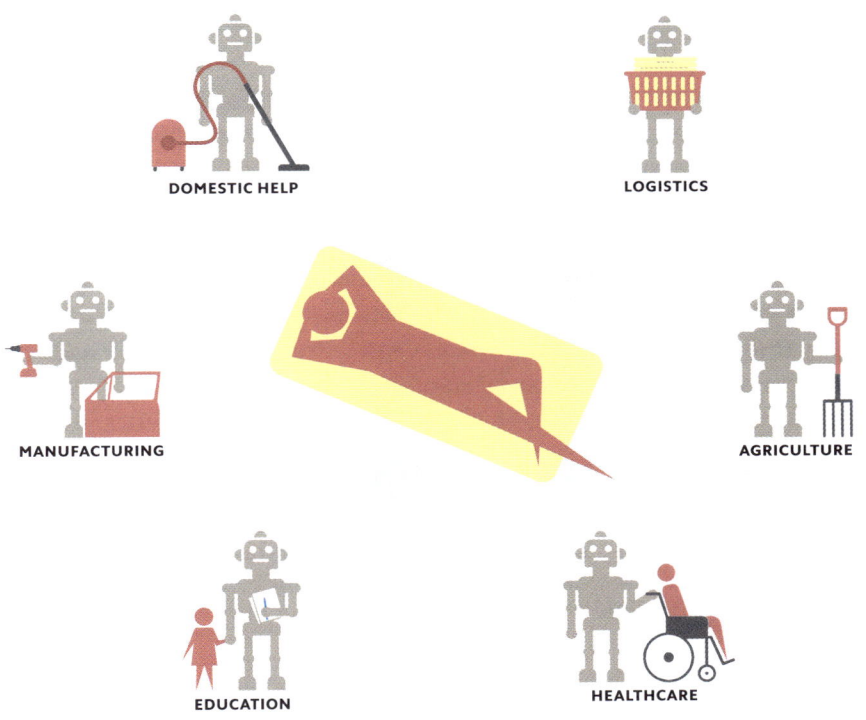

DOMESTIC HELP

LOGISTICS

MANUFACTURING

AGRICULTURE

EDUCATION

HEALTHCARE

# UNLIMITED REWARDS

Many AI researchers believe that AI will usher in a golden age for humanity – a time when machines will generate limitless abundance and prosperity. They argue that, with more powerful AIs doing all the work that humans used to do, people will finally be free to devote their time to leisure activities and to pursuing their personal dreams. At such a time, they claim, there will be no scarcity of resources, and so no crime, war, or injustice – and AIs will be able to help us to solve the world's remaining problems, from disease to global warming.

# INDEX

Page numbers in **bold** refer to main entries.

# ACKNOWLEDGMENTS

DK would like to thank the following for their help with this book: Vanessa Hamilton, Mark Lloyd, and Lee Riches for illustrations; Alexandra Beedon for proofreading; Helen Peters for the index; Jackets Coordinator Priyanka Sharma.

All images © Dorling Kindersley
For further information see:
www.dkimages.com

# S I M P L Y  EXPLAINED

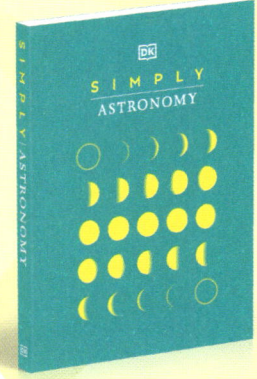

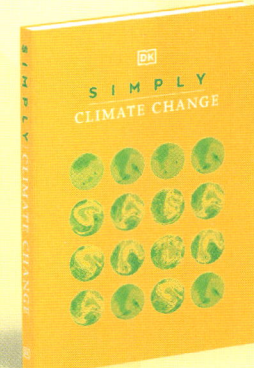

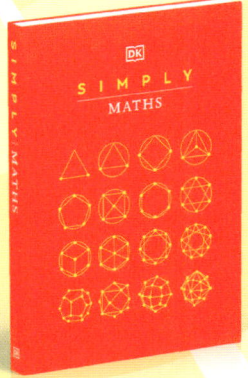

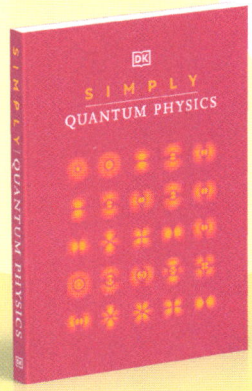

 For the curious